48 Topics in Current Chemistry
Fortschritte der chemischen Forschung

Stereochemistry II

In Memory of van't Hoff

Springer-Verlag
Berlin Heidelberg GmbH 19

This series presents critical reviews of the present position and future trends in modern chemical research. It is addressed to all research and industrial chemists who wish to keep abreast of advances in their subject.

As a rule, contributions are specially commissioned. The editors and publishers will, however, always be pleased to receive suggestions and supplementary information. Papers are accepted for "Topics in Current Chemistry" in either German or English.

Any volume of the series may be purchased separately.

DOI 10.1007/978-3-540-37943-0

Originally published by Springer-Verlag Berlin Heidelberg New York in 1974
MyCopy version of the original edition 1974

www.springer.com/mycopy

1852 J.H. van 't Hoff 1911

Contents

Springer-Verlag	D-6900 Heidelberg 1 · Postfach 1780 Telephone (06221) 49101 · Telex 04-61723 D-1000 Berlin 33 · Heidelberger Platz 3 Telephone (030) 822011 · Telex 01-83319
Springer-Verlag New York Inc.	New York, NY 10010 · 175, Fifth Avenue Telephone 673-2660

From van't Hoff to Unified Perspectives in Molecular Structure and Computer Oriented Representation

Dr. Johann Gasteiger, Paul D. Gillespie, Ph. D., Dr. Dieter Marquarding, and Prof. Dr. Ivar Ugi *

Laboratorium für Organische Chemie der Technischen Universität München

Contents

* and Department of Chemistry, Rice University, Houston, Texas

Abbreviations

AC-Matrix	**A**tom **c**onnectivity matrix
ASI	**A**tomic **s**equence **i**ndex
BE-Matrix	**B**ond and **e**lectron matrix
BPR	**B**erry **p**seudo**r**otation
CC-Matrix	**C**onfiguration and **c**onformation matrix
CIP-Rules	Sequence rules of **C**ahn, **I**ngold and **P**relog
EM	**E**nsemble of **m**olecules
FIEM	**F**amily of **i**someric **e**nsembles of **m**olecules
PI	**P**ermutation **i**somerization
RASI	**R**elative **a**tomic **s**equence **i**ndex (mostly of the α-atom of a ligand)
R-Matrix	**R**eaction matrix
TR	**T**urnstile **r**otation
$(TR)^2$	Double TR, *i.e.*, sequence of two TR with the same pair-trio combination without intermediate stop after the first TR.

Introduction

The interdependence of chemical constitution and the spatial arrangement of atoms in molecules suggests the simultaneous treatment of constitutional and stereochemical problems. Traditional separation of these latter aspects is more than likely due to the fact that the concepts and methodology of the two chemical disciplines differ enough to justify individual approaches.

A complete and detailed description of molecular structure includes statements concerning the metric coordinates[a] of atomic nuclei supplemented by electron density distribution data. Although a large amount of data is involved, such representation is not particularly suitable for the chemically relevant structural features of molecules.

The structure of a molecule is given by the three-dimensional distribution of atomic cores and valence electrons. This structure has been elucidated for many molecules with the use of X-ray or electron diffraction data. Chemical properties of molecules are observed under conditions which permit internal motions. Such observations yield views which may differ markedly as a function of time. Thus, observable properties are determined from equilibrated ensembles of species differing in geometry and energy.

This demands a representation invariant against changes of the latter features.

Thus, it behooves one to use problem oriented models for molecules, neglecting properties irrelevant for a given problem. Such models represent equivalence classes of molecules with a common characteristic feature. One must keep in mind, however, the limitations contained in the model's approximations.

For the majority of chemical problems, those models are the most useful which relate to the concept hierarchy of empirical formula, constitution, configuration and conformation:

Any empirical formula	⟹	All compounds
↓		
Same empirical formula different constitution	⟹	Constitutional isomers
↓		

[a] Cartesian or polar coordinates or the combination of bond lengths, angles and dihedral angles may be used in this context.

↓

Same constitution different configuration and/or conformation	⟹	Stereoisomers
↓		
Same configuration different conformation	⟹	Conformers

2. Constitution

In the first half of the last century, the concept of isomerism was discovered, *i.e.*, that there are distinguishable compounds with the same empirical formula. This led to the conclusion that within molecules, the atoms can be connected in different ways, hence, to the concept of constitution.

The chemical constitution of a molecule is given by a family of binary relations, namely, the covalently connected pairs of atoms.

The model of a covalent single bond is a pair of atom cores held together by a pair of electrons, with the electron density being notably different from zero everywhere between the cores. The number of electron pairs that participate in a bond is its formal bond order.

Some molecules exist where the bonding electrons cannot be assigned to atom pairs, but belong to more than two cores, *e.g.* in the polyboranes. In these cases the model concept of covalently bound atom pairs as a representation basis for chemical constitution using binary relations can be sustained by the assignment of fractional bond orders.

Sometimes it is advantageous to include the distribution of the free valence electrons in the constitution.

2.1. Formulas, Lists and Matrices

The constitution is customarily represented by constitutional formulas, *i.e.* by labelled graphs [1] whose nodes are the atoms, and whose connecting lines are the bonds. Nodes are labelled by chemical element symbols and can carry further symbols for electrons and electrical charges.

The information contained in a constitutional formula can also be given in terms of a connectivity list. Such a list tells which atom is connected to which (the formal bond orders are indicated, if needed). The statement that a C-atom is connected to another C-atom does not suffice, if there is more than one C—C-bond, but the bonded partners must be indicated individually. The connectivity list of *1* serves as an illustration.

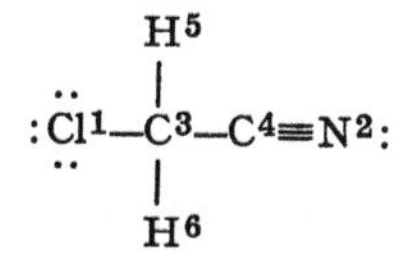

1

Cl1—C3, or just	1—3
N2—C4	2—4
C3 —C4	3—4
C3 —H5	3—5
C3 —H6	3—6

Fig. 1. Connectivity list of *1*

Rapid increase in the use of computers in Chemical Documentation and in the solution of other chemical problems lends increasing importance to connectivity lists and their corresponding atom connectivity matrices (AC-matrices)[2], as well as the associated bond and electron matrices (BE-matrices)[3].

$$E_1 = \begin{array}{c} \begin{matrix} 1 & 2 & 3 & 4 & 5 & 6 \end{matrix} \\ \begin{bmatrix} 6 & 0 & 1 & 0 & 0 & 0 \\ 0 & 4 & 0 & 3 & 0 & 0 \\ 1 & 0 & 0 & 1 & 1 & 1 \\ 0 & 3 & 1 & 0 & 0 & 0 \\ 0 & 0 & 1 & 0 & 0 & 0 \\ 0 & 0 & 1 & 0 & 0 & 0 \end{bmatrix} \end{array} \begin{array}{l} \\ 1,\ \mathrm{Cl} \\ 2,\ \mathrm{N} \\ 3,\ \mathrm{C} \\ 4,\ \mathrm{C} \\ 5,\ \mathrm{H} \\ 6,\ \mathrm{H} \end{array}$$

Fig. 2. BE-matrix of *1* (The AC-matrix is obtained by replacing the diagonal entries by the element symbols)

In the AC-matrices the off-diagonal entries e_{ij} are the formal bond orders between atom pairs (A_i, A_j). BE-matrices are obtained from the AC-matrices by augmentation with diagonal entries e_{ii} indicating the numbers of free valence electrons at the atoms A_i. The indices 1, ..., n can be assigned to the n atoms of a constitution in $n!$ different ways. Accordingly there are up to $n!$ different but equivalent connectivity lists, or AC- and BE-matrices, respectively. The direct identification and comparison of such representations is essential to Chemical Documentation. Only by uniquely assigning atomic indices can this be accomplished.

Such a unique indexing of the atoms affords unique connectivity lists, or matrices, respectively, which may be termed canonical representation[b] of chemical constitution.

---

[b] The canonical form of a mathematical expression is a standardized one, particularly well-suited for a given purpose.

2.2. Constitutional Symmetry

An indexing procedure must take into account constitutional symmetry, *i.e.* within the indexed set, the indices of constitutionally equivalent atoms (such as the three H-atoms of a methyl group) and bonds must be recognizable as equivalence classes with permutable numbers. A canonical connectivity list is invariant against permutations of constitutionally equivalent atom indices. It suffices for representing equivalence classes of constitutionally equivalent bonds (*e.g.* the C—H bonds of benzene) the use of a single symbol. This leads to abbreviated connectivity lists. The atoms of a chemical constitution are then equivalent if they belong to the same chemical element (and isotope) and, further, have the same set of connectivities and neighbors. The expression "constitutional symmetry" refers to the permutability of equivalently labelled atoms within a constitution.

For example, in benzene, all H- and C-atoms are constitutionally equivalent. In chlorobenzene, however, only the o- and m-atoms are pairwise equivalent. Note here that the individual resonance formulas of delocalized bond systems are not distinguishable.

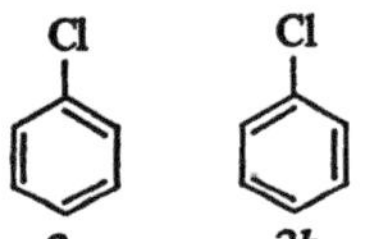

2a *2b*

NMR spectroscopy derives its usefulness, in large measure, from its ability to detect constitutional symmetry. If constitutional symmetry is not taken into account in synthetic design computer programs, a redundancy of pathways may result.

2.3. Sequentially Ordered Sets of Atoms

The sequence rules of Cahn, Ingold and Prelog (CIP rules) [4] have been defined for establishing sequential order in the ligand set of asymmetric C-atoms. Attempts to adopt these rules for indexing the atoms of a chemical constitution would lead to an indexing procedure requiring quite a cumbersome computer algorithm and attendant loss of efficiency. Since the CIP rules take into account formal bond orders, the assignment of atomic indices to delocalized bond systems by some modified set of the former would depend upon the choice of resonance formula (see *e.g.* the indexing of *2c* or *2d*).

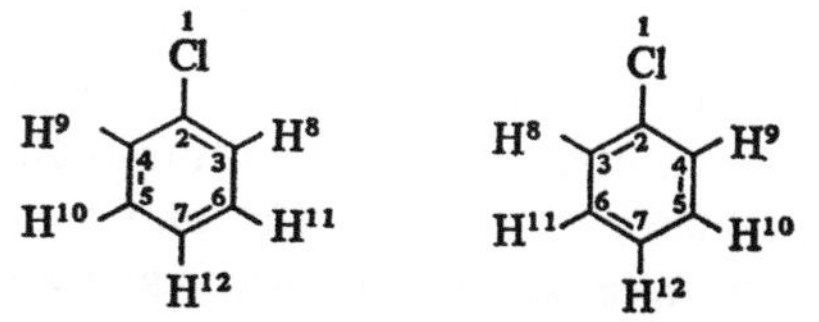

2c 2d

By the use of rules relying primarily upon atomic and coordination numbers it is possible to achieve an unambiguous ordering of atoms taking into account the bond orders indirectly via the coordination numbers. This extends to delocalized bond systems as well.

It is possible to consider first the coordination numbers, and then the atomic numbers, or to proceed in the reverse order. If there are more differences in atomic numbers than in coordination numbers the first mode is more effective, while the existence of many atoms of the same element with different coordination number favors the use of the second. Most atoms of a given chemical element have the same coordination number, particularly in organic compounds. If the H-atoms are, however, disregarded, one obtains constitutional formulas which can be indexed quite well by primary consideration of the degrees of the nodes[c], *i.e.*, the number of immediate covalent neighbors.

The Morgan algorithm [5] is a device used by the Chemical Abstracts Service for assigning indices to the nodes of constitutional formulas whose H-atoms have been omitted, *i.e.* their reduced graphs. If the indices of all atoms are needed, *e.g.* for the representation of stereochemistry, an additional procedure is needed for establishing those for the H-atoms. Constitutional symmetry is not indicated directly by Morgan indices.

A recently developed computer program [3f] for synthetic design on the basis of a mathematical model requires the transformation of arbitrary BE-matrices into the canonical forms. It was found that the atomic sequence indices (ASI) [6] are particularly well-suited for this purpose. The algorithm for assigning the ASI depends initially on atomic numbers, and secondly on coordination numbers. The following procedure for ordering atoms of a constitution is a translation of the ideas and criteria used in the ASI algorithm design.

a) The atoms are collected in chemical element equivalence classes by the atomic number; the higher the element number of an atom, the lower its index. In isotopically labelled compounds the mass number is also taken into account.

b) Atoms belonging to the same chemical element are distinguished according to the atomic numbers of their α-atoms, *i.e.*, their immediate

[c] One might as well remove all univalent atoms.

covalent neighbors. For each atom of an element equivalence class the atomic numbers of its α-atoms are written in a row in descending order from left to right. The number of rows resulting is the number of atoms in the element equivalence class, and the number of columns is the maximum coordination number of the chemical element considered. If an atom has fewer α-atoms than its maximum coordination number in the given compound, the missing α-atoms are represented by phantom atoms with atom numbers x, written in the empty spaces of the row on the right hand side of the known atomic numbers. Now, the entries of the first column are compared, with x being rated higher than any known atomic number. The atom corresponding to the row with highest first entry obtains the lowest available index of the element equivalence class. The atom belonging to the row with the second highest first entry receives the next index etc. Those indices of the element equivalence class which have not been assigned by the first column, are given to the remaining atom by analyzing the second, third, etc. columns in a similar position.

c) If the comparison of the α-atom set does not permit complete assignment of indices, the β-atoms are considered, *i.e.*, the atoms that are connected to a given atom via a sequence of two covalent bonds, and then the γ-atom (3-bond neighbor) etc. are taken into account in analogy to b), until all neighbors have been considered, or until all indices have been assigned. If there are rings the procedure ends at the latest when the cycle is completed.

d) Atoms which cannot be distinguished by a—c are constitutionally equivalent and correspond to an interval of equivalent indices.

The indices are permutable within any interval of indices belonging to constitutionally equivalent atoms, as long as the α-atoms of these atoms do not also belong to constitutionally equivalent classes. In the latter case, additional rules are needed for deciding between equivalent assignment of indices.

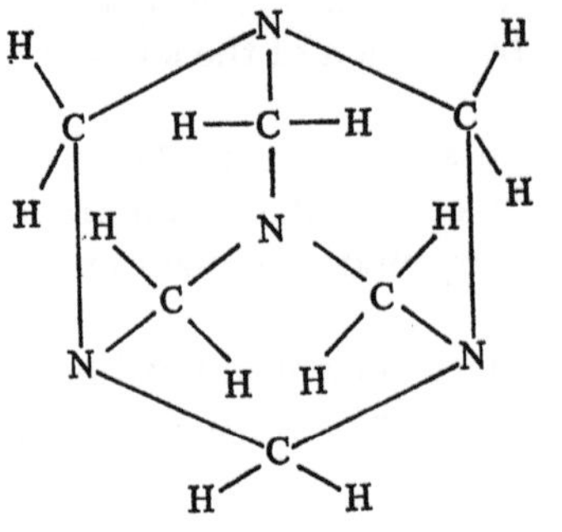

3a

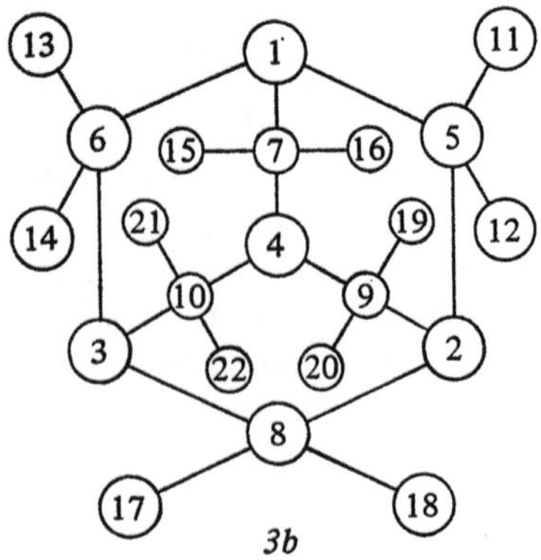

3b

In hexamethylene tetramine *3a* all atoms belonging to the same chemical element are constitutionally equivalent, and their indices belong to the following equivalence intervals

$$\begin{aligned} N&: \{1, \ldots, 4\} \\ C&: \{5, \ldots, 10\} \\ H&: \{11, \ldots, 22\} \end{aligned}$$

Within one of these intervals there is free choice of indices, but once this is decided the other assignments are no longer random. The arbitrary assignment of indices to the N-atoms determines the indices of the adjacent C-atoms and these in turn the order of the H-atoms.

It is further possible to describe a given constitution by mapping the indexed set of atoms onto the independently indexed graph. The indexed set of atoms must be partially symmetrized with regard to the element equivalence classes of the atoms and the indices of the graph with respect to its symmetry. That constitution is defined as the reference constitution in which the atomic indices a and the graph indices g match.

$$\begin{pmatrix} a \\ g \end{pmatrix}_{\text{ref.}} = \begin{pmatrix} 1, \ldots, n \\ 1, \ldots, n \end{pmatrix}$$

The permutational descriptor of the given chemical constitution is then a permutation of atom indices (or the inverse permutation of graph indices) which transforms the reference constitution into the considered constitution.

Such a procedure corresponding to the permutation-nomenclature system of configuration [7] is in accordance with the fact that a chemical constitution is a representative of a product $D, \times \ldots \times D_{\bar{a}}$ of the double cosets D_{δ} of pairs of subgroups symmetric groups $S_{n,\delta}$. The latter correspond to permutations of atoms of the same element and the symmetry of the adjacent subset of graph points whose degree is compatible with the coordination number of the element equivalence class of the atom.

Despite its mathematical clarity this permutation nomenclature is of no practical value, due to its cumbersome nature, requiring a rather complicated procedure for generating the graphs with standard indices.

3. Stereoisomers

It is customary to describe the essential spatial features of stereoisomers by their configurations and conformations.

The traditional limitation of the stereoisomer concept to isomeric compounds with different configurations stems from the idea that configurations are not interconvertible under the observations conditions.[8] A definition of stereoisomers on this basis is questionable as its application is then dependent upon observation conditions. It also fails to account for the fact that many mobile interconversions of configurations (*e.g.* configurations involving tricoordinate Nitrogen or pentacoordinate Phosphorus) are known, as well as thermally stable conformations (*e.g.* the atropisomers).

Present usage prefers considering any constitutionally equivalent molecules which are configurationally and/or conformationally distinguishable as stereoisomers. In the case of a mobile equilibrium of interconvertible conformations, that configurational data suffices which form class characteristics of the whole equilibrium ensemble of conformations. In the main, configurational data suffices, because it is possible to manufacture models of energetically preferred conformations by finding the corresponding flexional and torsional potential energy minimum. The present classification methods for stereoisomers are the product of a long historical development whose discussion is useful for understanding modern views.

3.1. Rigid Configurations

Rigid configurations are given by the distribution of a ligand set on the ligand sites of an achiral molecular skeleton[d] with a characteristic symmetry and coordination number, *i.e.*, the number of ligands that it can carry. Molecular skeletons can be monocentric, corresponding to the valence state of the atoms (*e.g.* sp^3, dsp^3, $d^2 sp^3$), or they can be polycentric, composed of monocentric units, such as in *4*, a skeleton that corresponds to cyclo-propane.

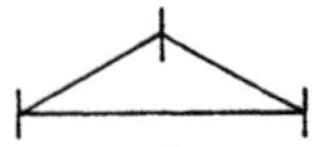

4

In this model, idealized skeletons are assumed, deformations of the skeleton by ligands being neglected. The ligands have a static or dynamic symmetry about their skeletal bond axes which is compatible with the skeletal symmetry. Here it should be noted that the conceptual dissection of molecules into skeleton and ligands has been a standard procedure developed by stereochemists quite some time ago.

---

[d] For reasons which are given below the present discussion is confined to configurations of achiral skeletons.

3.2. Ligands and Skeletons

It is not likely that a universal definition of skeleton and ligands will ever be generated. These might be concepts without definition, analogous to those of a *set* or a *point* in mathematics.

The following criteria are useful in the dissection of molecules and permit an assignment of the various parts of molecules to the skeleton or the ligand set, without pretending to define them.

a) Only one single or multiple covalent bond is dissected, in separating a ligand from the residual molecule.

b) Ligands have a symmetry about their skeletal bond compatible with the skeletal symmetry, *e.g.* cylindrically symmetric, or there is free rotation about this bond.

c) What remains after removal of all ligands is the skeleton.

If there are several ways of establishing the ligands and the skeleton according to the above criteria it is advantageous to take that skeleton that is common to as many of the molecules under consideration as possible.

Mapping the indexed ligand set of a configuration onto the indexed skeletal sites with due accounting for skeletal symmetry affords a consistent nomenclatural description as well as a procedure for enumerating the conceivable configurations of a skeletal class.

With more than one monocentric unit the overall configuration of molecules with inhibited internal rotations may also be determined by the dihedral angles of the bond system between monocentric units. The latter may no longer be treated as independent. A representation by polycentric configurations is advisable.

3.3. Conformations

The conformations of molecules are interconvertible by internal rotations about bond axes. A conformation of a molecule with n atoms can be represented by a family of

$$\binom{n}{4} \leqslant n(n-1)\,(n-2)\,(n-3)/4!$$

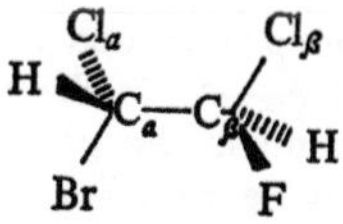

5

dihedral angles[e]. This type of representation contains sufficient information about the configuration. If the monocentric configurations are known, then $\binom{n}{2} \leqslant n(n-1)/2$ dihedral angles are adequate to describe the spatial arrangement, and often just their algebraic signs suffice.

For example, the representation of *5* requires information on the configuration about both C-atoms as well as the statement of at least one dihedral angle, *e.g.*, $\delta(Cl_\alpha{-}C_\alpha C_\beta{-}Cl_\beta) = -60°$. If a polycentric molecule corresponds to the equilibrium of an ensemble of conformations, one could describe it with the dihedral angles of all participating conformations, or neglect the conformational aspect and use the configurations as the characteristic class feature of the ensemble. Except for conformation analytic problems, the latter treatment suffices for the solution of chemical problems.

3.4. CC-Matrices

The essential stereochemical features of molecular systems with n atoms can be described by data on $<n$ monocentric configurations and $<n(n-1)/2$ dihedral angles which can be collected in C, an $n \times n$ configuration and conformation matrix (CC-matrix).

In a CC-matrix C the diagonal entries c_{ii} are the permutational or parity descriptors of configurations about the atoms A_i, and the off-diagonal entries c_{ij} are dihedral angles $A_k{-}A_i\,A_j{-}A_e$, or their algebraic signs. When dihedral angles $A_k{-}A_iA_j{-}A_l$ are stated, the monocentric valence skeletons are assumed to be known. Bond angles for monocentric skeletons follow from their coordination number and symmetry. The reference atoms A_k and A_l are those α-atoms of A_i and A_j having the

[e] A dihedral angle $\vartheta\ (A_1{-}A_2A_3{-}A_4) \leq 180°$ at a bond $A_2{-}A_3$ refers to a sequence of three bond directions

and is the angle between the projections of two bonds $A_1{-}A_2$ and $A_3{-}A_4$. The axis of the refence bond $A_2{-}A_3$ is the direction of the projection, and a plane perpendicular to the bond is the projection plane. Dihedral angles corresponding to a counterclockwise rotation of the bond which is closer to the observer such as the angle $\vartheta\ (A_1{-}A_2\ A_3{-}A_4)$ have a negative algebraic sign.

lowest indices. Entries $c_{ii} = 0$ and $c_{ij} = 0$ mean that no configuration or conformation data exist for A_i, or A_i—A_j, respectively. The CC-matrix of *6* illustrates this type of representation.

6

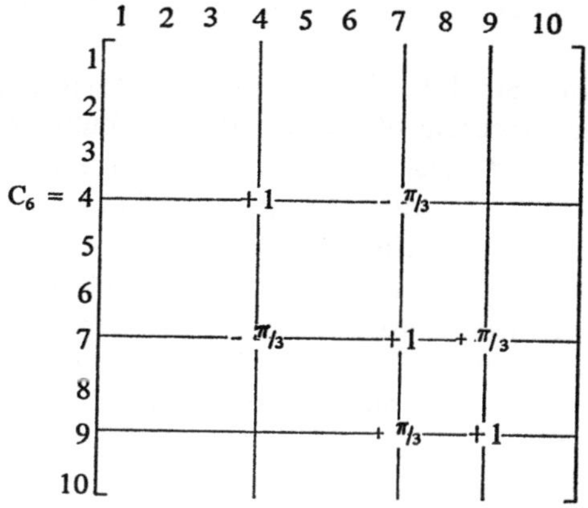

Fig. 3. CC-Matrix of *6*. Parity descriptors used for the diagonal entries are described in Section 5.6

4. Permutation Isomers and Chiral Configuration

It was not before 1937 that the classical study of Pólya [9] achieved the necessary critical analysis and generalization of the concept of configuration as given by van't Hoff. Pólya recognized that rigid configurations are distinguishable only if they are neither interconvertible by permutation of indistinguishable ligands, nor by rotations that belong to the skeletal symmetry. The widespread consequences of this fundamental study are receiving more and more attention of late.

Ruch [10a,b] was able to demonstrate that not all rigid chiral[f] configurations can be rigorously classified into right and left handed classes, as had been tacitly assumed, in analogy to asymmetric C-atoms. For many configurations belonging to Ruch's class B there is no chirality oriented nomenclature whose descriptors distinguish between the algebraic signs of chirality observations.

4.1. From Biot to van't Hoff and Beyond

In 1815 Biot [12] recognized that certain liquid organic compounds and also the solutions of some solid substances like saccharose, camphor, and tartaric acid are capable of rotating the plane of linearly polarized light. He ascribed this to some inherent property of the compound molecules. This initiated a development which led to the concept of stereochemistry.

Six years later Sir John Herschel [13] observed that the algebraic sign of the optical rotation of quartz crystals could be correlated with their shape, and the dextro- and levorotatory crystals look like mutual mirror images. This correlation was extended from crystals to molecules by Pasteur [14], who postulated that the spatial arrangement of atoms in the molecules is responsible for their optical activity, and the dextro- and levorotatory molecules are in a mirror image relation.

About the same time, Kekulé [15] discovered the quadrivalency of C-atoms laying the foundations for constitutional chemistry.

---

[f] Lord Kelvin [11a] recognized that the term "asymmetry" does not reflect the essential features, and he introduced the concept of *chiralty*. He defined a geometrical object as chiral, if it is not superimposable onto its mirror image by rigid motions (rotation and translation). Chirality requires the absence of symmetry elements of the second kind (σ- and S_n-operations) [11b]. In the gaseous or liquid state an optically active compound has always chiral molecules, but the reverse is not necessarily true.
Since the majority of organic molecules is conformationally and/or configurationally flexible, the consideration of a single rigid molecular model is not sufficient for the treatment of chemical chirality problems. As a rule it is necessary to take into account the equilibrium ensemble of the interconverting spatial arrangements. We call an ensemble achiral if it contains only achiral species (with an achiral rigid model) or if both enantiomers of the participating chiral species occur in equal amounts and form a "racemic" ensemble. If there exist chiral species without being compensated by their respective enantiomers the ensemble is termed "chiral". For chiral ensembles observations of chirality have a value different from zero unless the pertinent molecular chirality function has a zero for the chosen type of observation or the sum of the molecular chirality functions is zero for the given ensemble such as a suitable mixture of permutational isomers of allene derivatives [10b,c].

This led to the question of whether there is a rigid geometrical model with well-defined directions and angles for the spatial distribution of the four valencies of carbon, or if there is complete angular flexibility, without barriers. The answer came in 1874 when simultaneously van't Hoff [16] and Le Bel [17] made the statement that the four valencies of carbon point from the central C-atom towards the vertices of a regular tetrahedron. Furthermore, van't Hoff recognized that the so-called optical isomerism of carbon compounds is due to asymmetric carbon atoms, *i.e.*, carbon atoms which are connected to four different atoms or atom groups, and that only a tetrahedral arrangement of the valencies is compatible with the observed facts.

A group theoretical treatment shows that a monocentric tetracoordinate skeleton carrying four different ligands can lead to only two mirror-image like configurations, if the skeleton possesses tetrahedral T_d symmetry. In the absence of skeletal symmetry one could expect $4! = 24$ different tetracoordinate configurations. For completely flexible monocentric skeletons without flexion barriers and well-defined bond angles there would be no stereoisomers, only one observable configuration.

The impact of van't Hoff's concept of asymmetric C-atoms upon the thought of stereochemists was so strong that many configurational problems were treated on the basis of the model without questioning its pertinence for the given case.

For an asymmetric C-atom it is necessary that all four ligands be different, such that any odd number of their pairwise exchange produces distinguishable mirror images of themselves. The fact that these statements do not apply to configurations other than tetracoordinate ones with T_d skeletal symmetry did not receive due attention for a long time.

4.2. Transitivity Prerequisites of Homochirality and Heterochirality

The concepts of homochirality and heterochirality [10a,b,11a] are useful in the comparison of configurations belonging to Ruch's class A [10]. The partitioning of chiral compounds of Ruch's class A into two antipodal classes requires some well-defined criterion of configurational similarity sufficient to decide whether two configurations belong to the same class (they are homochiral) or not (they are heterochiral). Unfortunately, this general criterion has yet to be found. The decision whether a configuration is similar to another one or to the mirror image of the latter if often arbitrary.

The concepts of homo- and heterochirality play an important role in the discussion of the stereochemical aspects of reaction mechanisms. In many textbooks [18] one finds the statement that the S_N2 mechanism is characterized by the fact that it occurs with configurational inversion.

The term "Walden inversion" also reflects this viewpoint. With the exception of substitution reactions where entering and leaving ligands are chemically indistinguishable the products of S_N2 reactions are by no means the mirror images of the starting materials. Rather they are merely more similar to the mirror images of the starting materials than to these themselves. If one considers the entering and leaving groups as equivalent, then the products and starting materials of S_N2 are heterochiral. Conversely, a homochiral relation exists between the products and starting materials of "retentive" substitution reactions.

When using these concepts one has to keep in mind that homochirality by chemical similarity is not always a transitive relation and is thus not suitable for establishing equivalence.

For example, the homochirality by chemical similarity of *7a* and *7b*, as well as *7a* and *7c* does not imply that *7b* and *7c* are also homochiral.

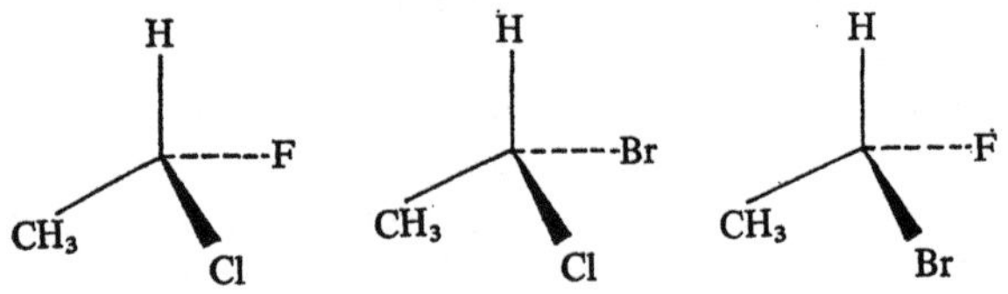

7a *7b* *7c*

Homochirality is an equivalence relation only when the ligands of all considered configurations of a skeletal class are ordered sequentially and is stated on the basis of this order.

Ordering of the ligands of *7a–c* by the CIP or ASI rules (Br > Cl > F > CH_3 > H) establishes homochirality for *a* and *c*, but not for the pairs *a*, *b* and *b*, *c*.

4.3. Superposition of Monocentric Configurations

The distinction between the various configurations of monocentric, tetrahedral molecules is dependent on the ability to differentiate between all four ligands. In systems with more than one asymmetric C-atom these can be regarded as the chiral subunits of the total configuration. If there is sufficient flexibility one can expect 2^n configurations with n different asymmetric C-atoms. If there are asymmetric C-atoms of the same kind, as in the isomers of tartaric acid, a lower number of distinguishable configurations is encountered.

The n asymmetric C-atoms of molecules with rigid polycentric skeletons cannot be treated as independent chiral subunits of the overall configuration. Further, the number of conceivable configurations is no longer 2^n. For example, a rigid ethane skeleton with D_{3d} symmetry and six distinguishable ligands possesses 12 stereoisomers, not $2^2=4$, as could be expected with two asymmetric C-atoms. For a compound of type *8* without an asymmetric C-atom, there should be $2^0=1$ stereoisomer, however, the number of stereoisomers *8a—c* is three. Of these one, *8c*, is achiral and two, *8a* and *8b*, form a pair of chiral antipodes. This

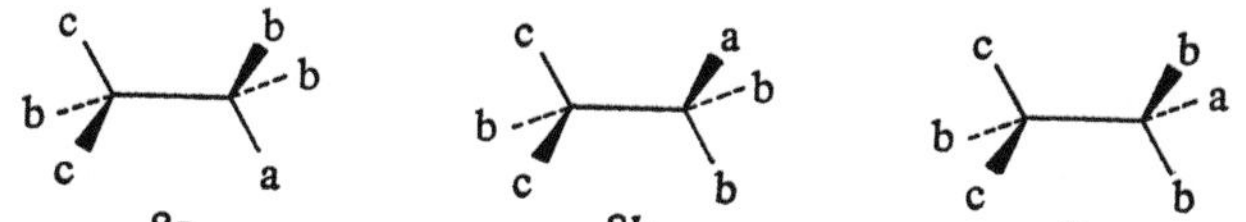

8a *8b* *8c*

example also demonstrates that with rigid skeletal models one encounters chirality without asymmetric C-atoms.

van't Hoff [19] already pointed out that asymmetric C-atoms are not a sufficient condition for optical activity, as is demonstrated by the examples of cis-cyclopropane dicarboxyclic acid *9* and meso-tartaric acid *10*.

van't Hoff's interpretation of the achirality (in his terms optical inactivity) of the meso-forms by stating that the contributions of both asymmetric C-atoms cancel each other is true in a geometrical sense only for *9*.

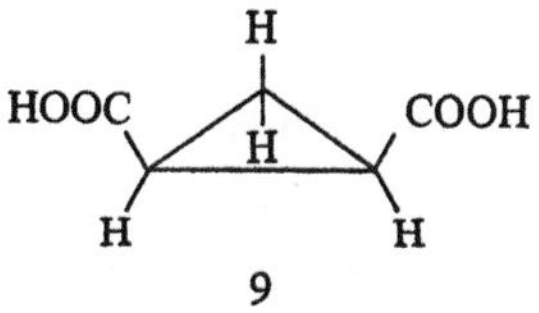

9

The latter exists as only one achiral polycentric configuration. For systems like *10* there are not only achiral rigid models *10b* and *10d* but also chiral ones, *10a* and $\overline{10a}$, as well as *10c* and $\overline{10c}$ where the observable achirality must be interpreted by looking at the whole ensemble of conformations. Such systems are achiral, because the enantiomeric conformations are mutually interconvertible via achiral conformations and occur at equal equilibrium concentrations.
The existence of achiral conformations in the ensemble is no prerequisite for its achirality. K. Mislow *et al.*[20a)] synthesized an achiral diphenyl derivative with orthogonal benzene rings which carry in their p-positions chiral residues with the same chemical constitution but opposite configuration and observed achirality although none of the conformations is itself achiral. Here achirality is attributed to the fact that the molecule

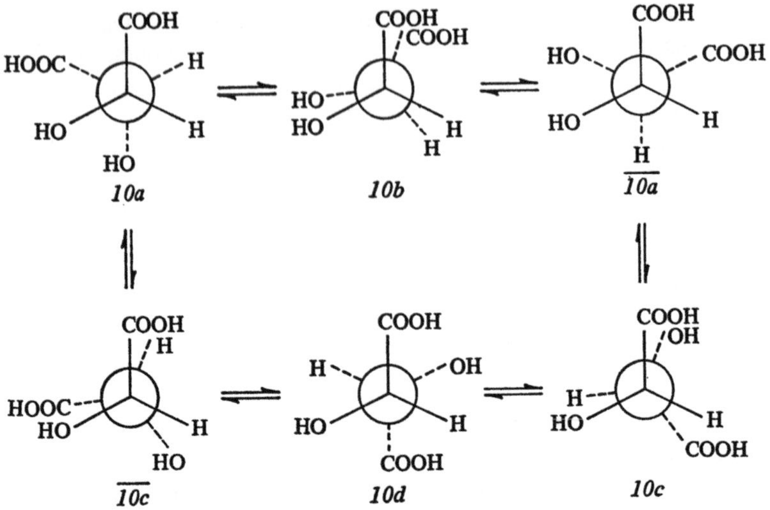

and its mirror image can be made superimposable by internal rotation of the diphenyl moiety versus the chiral substituents [20b,8a], *i.e.*, the achirality is due to the racemic nature of the ensemle of the chiral conformations. This example demonstrates that the customary treatment of stereochemical problems on the basis of rigid model point group symmetries is not always adequate.

4.4. Aschan's Criticism of van't Hoff

The fact that the existence of asymmetric C-atoms is *neither necessary nor sufficient* for chirality was pointed out in 1902 by Aschan [21] in his criticism of van't Hoff's concept. He used the examples of meso-tartaric acid *10* and the achiral trioxiglutonic acid *11* to demonstrate that asymmetric C-atoms are not sufficient for chirality, and example *12* shows that they are also not necessary.

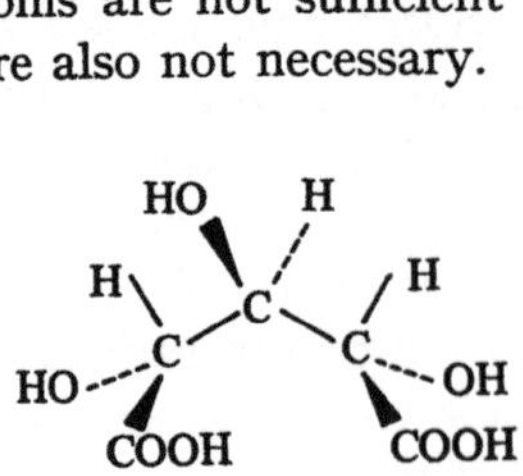

11

12

The critical argument of Aschan that asymmetric C-atoms are not necessary for chirality, *e.g.*, of *12*, does, in fact, not pertain to van't Hoff's concept. This latter involves only the assumption that the combination of chiral units leads to chiral molecules and is deliberately confined to systems of the latter kind. Chirality of compounds similar to *12* are not within the scope of van't Hoff's postulation, since the latter applies only to acyclic and monocyclic carbon compounds.

If the problem were to partition a set of carbon compounds into two equivalence classes, of which one contains only chiral molecules and the other one only achiral ones, it could not be solved with the criterion of asymmetric C-atoms. In the first case, one would assign meso-forms like *9* and compounds with pseudo-asymmetric [22] C-atoms, such as *11*, to the chiral equivalence class, and in the second, chiral molecules like *12* would remain in the achiral subset. However, the latter class would be devoid of chiral molecules, if the compounds under consideration have been confined to molecules with free rotation about all C—C bonds.

4.5. Cyclic and Polycyclic Configurations

All difficulties arising from an uncritical use of van't Hoff's concept are due to the treatment of the chirality of compounds with rigid polycentric skeletons on the basis of asymmetric C-atoms. For example, the chirality of the norbornane derivatives, *i.e.*, *13*, has been attributed to its asymmetric bridgehead C-atoms [20c,23].

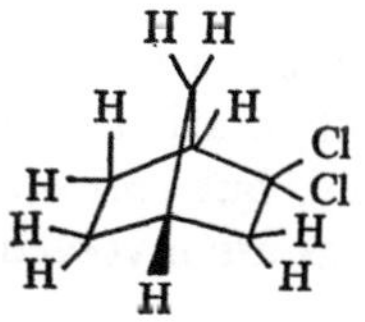

13

The replacement of the bridgehead atoms by planar tricoordinate skeletal units [g] should then lead to an achiral molecule. As is seen from example *14* this is not true, *i.e.*,

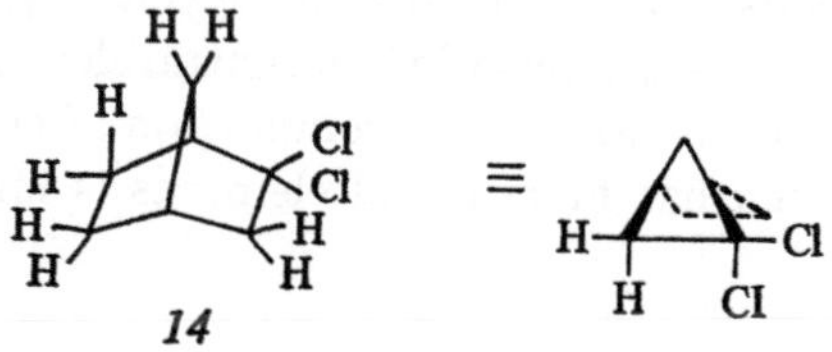

14

[g] Planar tricoordinate configurations cannot be chiral, because they contain a skeletal plane of reflection.

asymmetric C-atoms are not responsible for chirality, In this context a common error in many textbooks [23] may be noted.

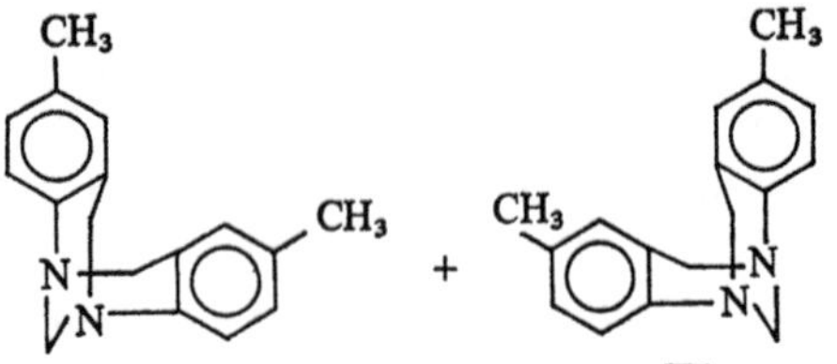

15 $\overline{15}$

The fact that Tröger's base [24] is a racemate which can be resolved [25] into its antipodes *15* and $\overline{15}$ is no proof for the asymmetry of the bridgehead N-atoms. Of the two conceivable arrangements in which the N-atoms and their direct neighbors (α-atoms) are coplanar, only that case with all atoms except the aliphatic hydrogens coplanar leads to racemization after inversion of the configuration of the N-atoms. However, the molecule in a situation similar to *14* is still chiral, even after inverting the configurations of both N-atoms which leads to an arrangement similar to *16a*.

One might question the description of polycyclic systems using only the configuration of their asymmetric C-atoms without adequate dihedral angle data. This is illustrated by the fact that the inversion of the local monocentric configuration does not necessarily lead to the mirror image of the molecule, but corresponds to an in-out-isomerization[h] if sufficiently large rings are involved [26].

4.6. Chirality Elements

The representation of stereochemical features by a polycentric configuration contains the monocentric configurations and essential information on the conformations. Use of the polycentric skeleton concept is particularly advantageous for cyclic and polycyclic systems. Sometimes, the asymmetric C-atoms are regarded as central elements of chirality[i] [4], and their enumeration is used in the comparison of chiral systems when the question of which has more chiral elements is important.

---

h) In the case of in-out-isomerism there are more than 2^n stereoisomers with n asymmetric carbon atoms, even if one disregards systems with crossed bridge.

i) Configurations cannot be unambiguously classified by using the concept of central, axial and planar elements of chirality. [4]

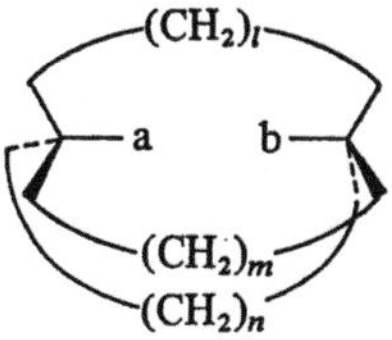

16a

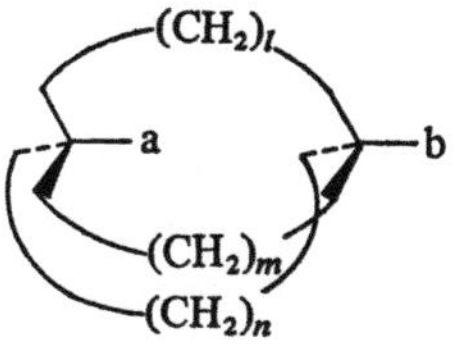

16b

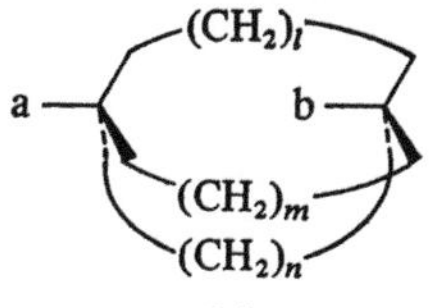

16c

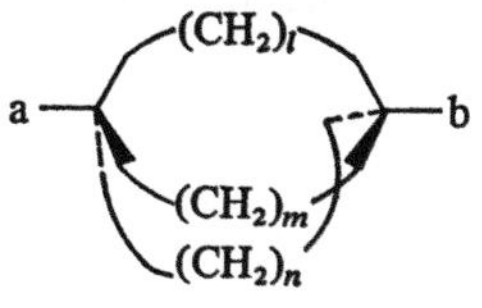

16d

The above discussion indicates that this viewpoint is not justified, even in general, for molecules with uninhibited internal rotation.

The use of the previously defined procedure (see Section 3.2) for dissecting a molecule into its skeleton and ligand set permits one to enumerate the elements of chirality for monocentric configurations with higher coordination numbers.

If a molecule can be dissected in various different ways into an achiral skeleton and ligands with an appropriate symmetry about the skeletal bond axes (see Section 3.2) all alternatives have to be considered.

For each symmetry element of the second kind (planes of reflection and improper axes of rotation) one counts according to Eq. (1) the pairs of distinguishable ligands at ligand sites which are superimposable by symmetry operations of the second kind.

$$u = \frac{1}{2}(m - s - t) \tag{1}$$

m = total number of ligands,

s = number of indistinguishable ligands which are superimposable by symmetry operations of the second kind,

t = number of ligands at skeletal sites which lie on a symmetry fix of the second kind (plane of reflection or improper axis of rotation),

u = number of pairs of distinguishable ligands which occupy skeletal sites superimposable by symmetry operations of the second kind.

For the individual symmetry elements of the second kind different values for u may result. The lowest value of u, called u_{min}, is the number of chirality elements of the given rigid model of the molecule. These elements of chirality are responsible for the non-appearances of the skeletal symmetries of the second kind in the observable molecular symmetry. u_{min} is the minimum number of pairwise ligand exchanges at equivalent skeletal sites needed for the transformation of a molecule into its mirror image.

For example, in the tetracoordinate configuration *17* with a T_d skeletal symmetry we have $u = u_{min} = 1$ for all planes of reflection. For all chiral molecules of Ruch's class A the same is true.

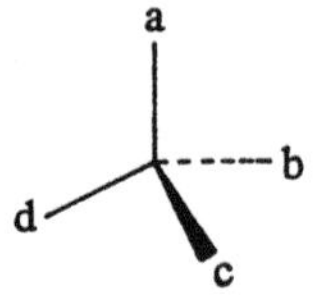

17

In the enumeration of chirality elements of flexible molecules all arrangements are taken into account which are permitted by the given constraints under the observation conditions. Here, one must always assume a rigid skeletal model and freely rotating ligands[j]. That arrangement for which the lowest number of chirality elements is found equal zero determines the number of chirality elements for the whole ensemble.

The equilibrium ensemble of the configurations of meso-tartaric acid *10a—c* has $u_{min} = 0$. From the definition of u_{min} follows immediately that for all achiral molecules $u_{min} = 0$ holds. This applies also to those cases where asymmetric C-atoms are present but cancel each other because of overall molecular symmetry.

Meso-tartaric acid *10* has no element of chirality, because there is a conceivable conformation *10b* with a mirror plane perpendicular to the central C—C-bond.

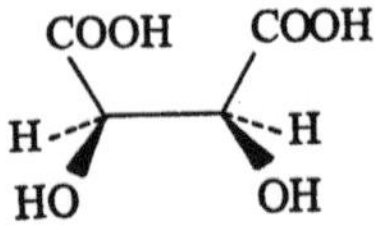

10b

[j]) Deformations of the skeleton by different ligands are disregarded. The necessity for this assumption was already recognized by van't Hoff.

If in *10* the equivalence of one pair of ligands is destroyed (*e.g.*, by the esterification of one of the carboxyl groups, to form *18*) it results in $u_{min} = 1$. This means that only one pair of ligands (CO_2H and CO_2R) must be exchanged in order to obtain the mirror image of *18*; here the conformation *18a* is considered if the exchange refers to a reflection, and *18b* if a center of inversion is the reference.

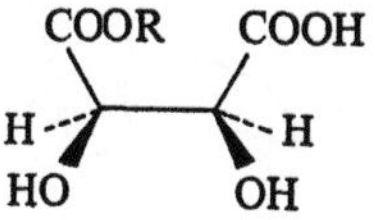

18a

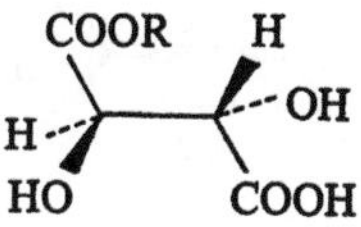

18b

According to the classical concept of van't Hoff one would need two operations for the conversion of *18* into its mirror image, namely the inversion of both asymmetric C-atoms. These two examples demonstrate the advantages of the present procedure for enumerating the chirality elements of molecules.

Execution of the number of pairwise ligand exchanges, as given by u_{min}, can lead to the mirror image via permutation isomers (see below) which are constitutionally different from the considered system. The conversion of *13* into its antipode $\overline{13}$ via *19* or $\overline{19}$ serves as an illustration.

Molecules exist whose chirality elements can be enumerated according to the outlined procedure, but for which one finds a lower number of chirality elements by a different procedure. For example, the dissection of *20* and *21* into achiral skeletons and sets of achiral ligands (H, H, H, a, b, c) leads to three chirality elements by regarding the achiral

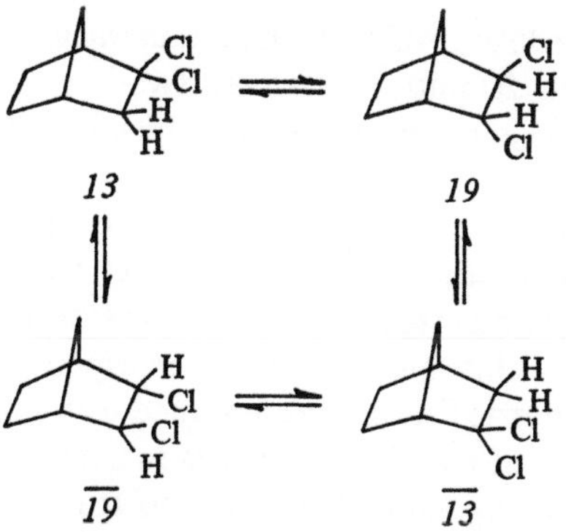

methyl and the chiral substituted methyl groups as ligands (*i.e.*, one local chirality element with regard to the skeleton, add another local one in the chiral ligand). These enumeration alternatives correspond to the fact that *20* and *21*, respectively, can be transferred into their mirror images either by stepwise exchange of a, b and c with the H-atoms of the methyl group, or by exchange of the unsubstituted and the

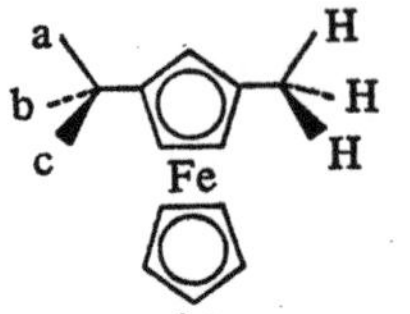

20

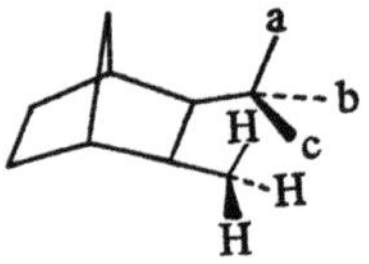

21

substituted methyl groups in combination with an inversion of the latter by a pairwise exchange of ligands. The latter procedure is always possible when a chiral molecule is dissectable into an achiral skeleton and a ligand set containing chiral ligands.

Hereby, it is also possible to enumerate the chirality elements of a variety of chiral molecules which cannot be treated according to Eq. (1) because they cannot be dissected into an achiral skeleton and a set of achiral ligands.

The dissection of *22* leads on one hand to a *chiral skeleton* and a set of *achiral ligands*, and on the other hand to an *achiral skeleton* and a set of ligands that contains a *chiral ligand*.

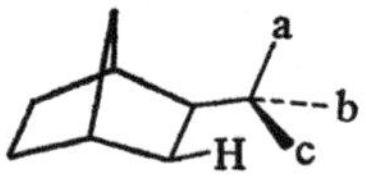

22

The latter procedure results in the enumeration of chirality elements for *22* in analogy to *20* and *21*, namely two.

In such cases the enumeration of chirality elements is generally achieved stepwise.

In one step, the chirality elements are counted with reference to the skeleton, and in another, the local chirality elements of the separated chiral ligands are enumerated. The resulting sum is the total number of chirality elements. This enumeration corresponds to the number of independent asymmetric C-atoms as assumed by van't Hoff. Furthermore, it yields the number of chirality elements in systems with "chiral units" of a different nature, *e.g.* some belonging to Ruch's class B. This simply requires the additional analysis of the chiral ligand obtained from Eq. (1).

Chirality element enumeration is essential for the classification of stereoselective reactions [27]. For instance, in order to distinguish an asymmetrically induced synthesis from other reactions whose stereoselectivity is also due to a chiral reference system, one must compare the number of chirality elements in the starting materials and the products.

5. Stereochemical Nomenclature Systems

The currently used stereochemical nomenclature systems for configurations with four or more ligands are chirality oriented, refering to rigid configurations, or their monocentric subunits. The preceding discussion demonstrates, however, that in many cases it is preferable to use a polycentric representation.

5.1. Ambiguities of the Classical D,L-Nomenclature

In 1891, Fischer proposed [28] describing molecules with asymmetric C-atoms with their two-dimensional projections. This projection is illustrated with the example of D-glyceraldehyde *23*.

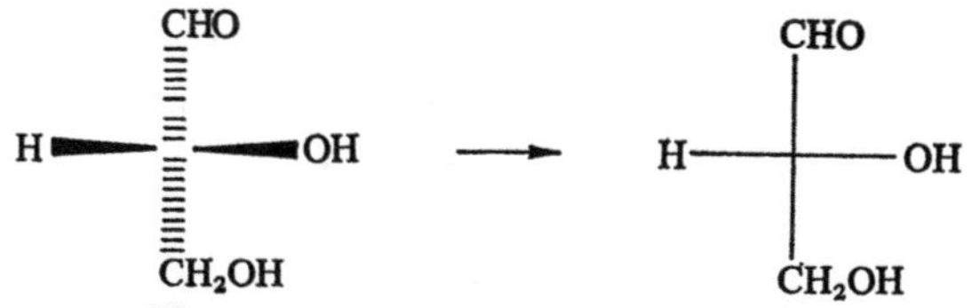

23a *23b*

Among the glyceraldehyde enantiomers, that one is called the D-enantiomer whose OH-group is on the right hand side of this projection, and its mirror image being the L-enantiomer. D-glyceraldehyde serves as the reference for the nomenclatural classification [8] of the configuration of other compounds. The unambiguous correlation of a given configuration to D-glyceraldehyde amounts to establishing a homochiral or heterochiral relationship. Without sequential order of all ligands involved, however, this correlation is not possible in a consistent manner (see Section 4.2).

5.2. CIP-Rules and *(R,S)*-Nomenclature

Cahn, Ingold and Prelog [4] recognized that the sequential ordering of ligand sets is an essential prerequisite for the nomenclatural classification of configurations. Their resulting sequence rules (CIP rules) generated the first unambiguous, nomenclatural classification for configurations.

According to the CIP rules ligands of monocentric configurations are indexed initially with the aid of the atomic numbers of the ligand atom, beginning with their α-atoms and proceeding from these outbond as far as is necessary for establishing pairwise greater-than relations. The elements of a set can be completely ordered by a family of pairwise greater-than relations. That ligand obtains the lowest index, or highest priority whose α-atom has the highest atomic number. If the α-atoms do not all differ in atomic number, the outbond β-atoms are compared etc. until a discrimination is possible. Multiply bonded atoms are taken into account as many times as is indicated by the bond order. A doubly bonded atom is rated lower than two singly bonded atoms of the same kind.

The currently used (*R*,*S*)-nomenclature of asymmetric C-atoms and similar tetracoordinate configurations is based upon the sequential order of the ligands derived from the CIP rules. If one orients an asymmetric C-atom in such a manner that the fourth ligand points backwards, the indices of the first three ligands of an (*R*)-configuration increase in a clockwise pattern, and counterclockwise for an (*S*)-configuration.

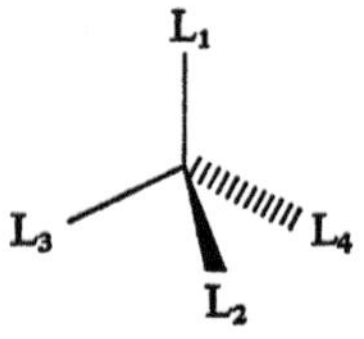

24

The (*R*,*S*)-nomenclature still reminds the user of the right and left handed helical pattern arising from Fresnel's [29] interpretation of optical activity. These patterns are characterized by the combination of a translational and a rotational direction[k]. The T_d skeletal symmetry of tetracoordinate systems submits itself to the pictorial models not applicable to other configurational types. The CIP rules may as well be used to define a configurational nomenclature on the basis of the Fischer projection. If one specified that in such a projection of an (*R*)-

[k] Kinetic chirality corresponds to this picture, such as the chirality of a moving particle with angular and translational momentum.

configuration the first ligand is on the $+Y$-axis and the other three follow a clockwise pattern. Such a nomenclature would be a another representation for the (R,S)-system.

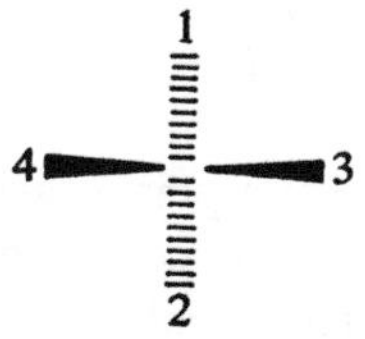

25a

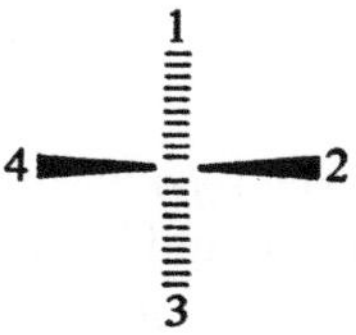

25b

5.3. Classical and Modern Nomenclature Systems of Bicentric Configurations

The *threo-erythro*-nomenclature of bicentric diastereomers with two adjacent asymmetric central atoms and free rotation about the connecting bond suffers from the same deficiencies as the D,L-nomenclature. Here again, well-defined descriptor assignment can be achieved by the use of a ligand sequence rule. The *p,n*-nomenclature of bicentric configurations has been derived from the (R,S)-nomenclature and avoids ambiguity by its use of an ordered ligand set. [30]

The classical *cis-trans*-nomenclature[1] of planar bicentric rigid configurations fails to consistently provide an unambiguous assignment of descriptors, without adequate ordering of ligands. Therefore it is gradually being replaced by the E,Z-nomenclature [31] which is a CIP sequential *cis-trans*-nomenclature.

5.4. Permutation Isomers and Index Mappings

A universal configurational nomenclature must always afford a one-to-one mapping of descriptors and configurations. If possible, such a system should be easy to use and simply adapted for application by computers. Further, the interconversion between various nomenclatures would be greatly facilitated, if the major existent systems were contained as subsets within a single system. A nomenclature satisfying these conditions can be derived from an analysis of the mapping of ordered ligand sets of permutation isomers onto their indexed skeletal sites.

---

[1] *Cis-trans*-isomers as well as the *atrop*-isomers and the chiral allene derivatives can be represented as polycentric configurations, or by a pair of monocentric configuration and a dihedral angle.

Mapping of the ligand indices $l=\{1...m\}$ of permutation isomers onto their skeletal indices $s=\{1, ..., m\}$ indicates which ligand occupies which skeletal site. The standard mapping with matching pairs of indices

$$\begin{pmatrix} l \\ s \end{pmatrix}_{\text{ref,}} = \begin{Bmatrix} 1, \ldots, m \\ 1, \ldots, m \end{Bmatrix}$$

corresponds to the *reference isomer* of a class of permutation isomer with a common skeleton and ligand set. In the reference isomer (or configuration), ligand number one occupies skeletal site number one, etc. The other isomers are obtained from the reference isomer by ligand permutations.

5.4.1. CIP Ligand Indices and RASI

For the assignment of ligand indices, the ligand set is sequentially ordered. Use of the CIP rules for the ordering has the advantage that one obtains descriptors corresponding to those of the (R,S), as well as the E,Z- and p-n-nomenclatures. It is desirable to have a common indexing system for the simultaneous representation of the constitutional and stereochemical features of molecules. Since the CIP rules have certain drawbacks in numbering the atoms of a chemical constitution (see Section 2.3) they are not suitable for a unified indexing system.

The atoms of a given configuration which are directly connected with a skeleton are the α-atoms of the ligands. Their ASI can be used to indicate which ligand is connected to which skeletal site, without the need of defining the ligands explicitly. The α-atoms of indistinguishable ligands are constitutionally equivalent and their ASI belong to equivalence intervals. Accordingly the ASI indicate directly where needed whether configurations are distinguishable or not.

The ASI as such are suitable for the representation of configurations. However, in the description of configurations only the relative numerical order of the ASI is relevant, not their absolute values. Let B be an ordered subset of m atoms belonging to the set A of the n atoms of a constitution ($B \subset A$, $m \leq n$). Then the indices $1,...,m$ of B are its *relative atomic sequence indices.* (RASI)

If the subset B to which the RASI refer contains only the α-atoms of a monocentric configuration, then the pairwise greater than relation of any ASI or RASI refers only to the respective ligands and is independent of all other ligands. In this case the RASI correspond to a true ligand sequence rule and are in most cases in agreement with the CIP indices of ligands. A ligand set containing indistinguishable ligands can be indexed in a variety of ways which differ by their index permutations.

In the representation of configurations by a mapping of ligand indices one must take into account the fact that in ligand sets containg indistinguishable ligands the respective permutational symmetry of the ligand set gives rise to multiplicity of mappings. For example, in the ligand set

$$\mathcal{L} = \{Br, Cl, H, H\}$$

any one of the two H-atoms would be number 3 or 4 according to the CIP rules, and two equivalence assignments of indices would result:

$$\{1,2,3,4\} \text{ and } \{1\,2\,4\,3\}.$$

5.4.2. Skeletal Indices and Symmetry

Rigid skeletal models can be indexed in any arbitrary manner but once an assignment of skeletal indices has been made, it must be retained.

Skeletal symmetry leads to various equivalent index assignments differing by index permutations representing the rotations belonging to the skeletal symmetry. The following indexed pentacoordinate trigonal bipyramidal rigid skeletal models *26a—c* belong to the six equivalent indexings which are interconvertible by rotations about two and threefold

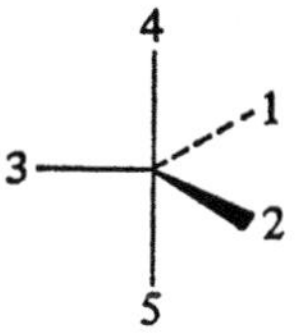

26a

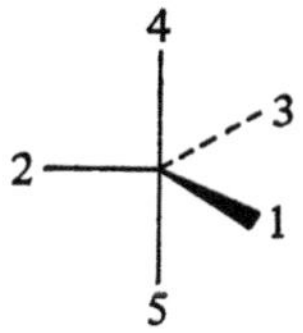

26b

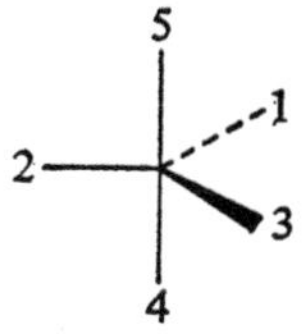

26c

rotation axes of the skeletal D_{3h} symmetry [32]. Let $\{1,2,3,4,5\}$ be the indexing of *26a*. The permutation (1 2 3) which corresponds to a rotation about the threefold apical axis[m] leads to $(1\,2\,3)\ \{1,2,3,4,5\} = \{3,1,2,4,5\}$ as the index assignment of *26b*.

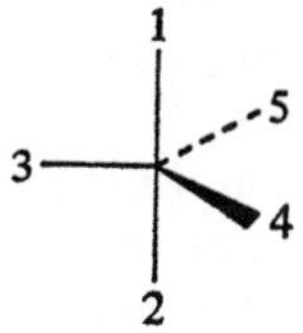

26d

If the flexion barriers of monocentric skeletons are such that the exchange of ligands occurs only by certain mechanisms, the number of

---

m) The permutation operator (1 2 3) indicates: the 1 replaces 2, the 2 replaces 3, and the 3 replaces 1.

observable dynamic configurations depends upon the ligand set and the characteristic features of the rigid skeletal models (its coordination number and symmetry). This physical description is represented by the number of disjoint parts contained in the graph of the mechanism which corresponds to the number of distinct configurations. [32c]

The pentacoordinate phosphorus compounds with a trigonal bipyramidal skeleton and five different ligands can exist in 20 different configurations. Their interconversions by the BPR, or TR mechanism [32], whose graphs have no disjoint parts, yields only one observable dynamic configuration, *i.e.*, all rigid configurations interconvert, whereas the TR [2] mechanism with a graph consisting of two disjoint parts corresponds to two distinct dynamic configurations.

The rigid model skeletal symmetry of flexible configurations and the skeletal flexibility of flexible configurations can be combined into the dynamic skeletal symmetry. The latter cannot always be represented by a Longuet-Higgins group.[33] If for example skeletal arrangements are interconverted by BPR, say *26a* and *26b* into *26d*, then the permutation of skeletal sites according to (1 4 2 5) is to be taken into account by considering the indexings {1,2,3,4,5} and (1 4 2 5) {1 2 3 4 5} = {5,4,3,1,2} as equivalent.

5.5. Permutation Descriptors

As stated earlier, ligand permutations by which a given configuration is obtained from a reference isomer are accomplished by a permutation of ligand indices. These permutations transform the mapping $\binom{l}{s}_{\text{ref.}}$ into the $\binom{l}{s}_{\text{conf.}}$ and can be further employed as a descriptor for the latter. The action of the inverse $P_s = P_l^{-1}$ of a ligand index permutation P_l upon the skeletal indices of $\binom{l}{s}_{\text{ref.}}$ has the same effect as the action of P_l upon the ligand indices. Thus, P_l and P_s are equivalent as permutation descriptor of configurations.

In the case of permutation isomers with some indistinguishable ligands and/or with skeletal symmetry there are different but equivalent ligand permutations leading to the reference isomer of the configuration. Among the numerically ordered equivalent ligand index permutations that one is chosen for the descriptor which contains the lowest number of transpositions[n].

---

[n] The inverse of a permutation will "undo" the effect of a permutation and is obtained from it by reversing the order of symbols, *e.g.* $(1\ 2\ 3\ 4)^{-1} = (4\ 3\ 2\ 1) = (1\ 4\ 3\ 2)$.

It is a particular advantage of the permutation descriptors that they do not only represent configurations, but provide also a basis for the representation of configuration interconversions. The permutational isomerization (PI) of a configuration A into a configuration B is described by the isomerizer $I_{AB} = P_B \cdot P_A^{-1}$, with P_A and P_B being the permutation descriptors of A and B.

The coset and Wigner subclass structures of the symmetric groups S_m permits to classify the regular PI[o] of a given skeletal class of configurations in terms of processes and mechanisms [32a,c].

Each isomerizer I represents a distinct PI. Those PI whose pathways of motions are equivalent by skeletal symmetry belong to the same mechanism, *e.g.* the BPR, or TR. Thus, a mechanism is an equivalence class of PI where the pathway of motion is the defining equivalence relation. Let $\hat{G}$ be a subgroup of S_m which represents the full skeletal symmetry, then those isomerizers $I \in S_m$ conjugate to a given $I' \in S_m$ (with $I' = \hat{G}I\hat{G}^{-1}$) form a subclass of conjugate elements and belong to the same mechanism.

PI, regardless of their mechanism, belong to the same process if they yield the same product from a given initial configuration Processes correspond to the cosets of G (with $G \subset \hat{G}$) representing skeletal automophisms, such as the rotations of the skeletal symmetry [34].

5.6. Parity Descriptors

Configurations of Ruch's class A occur pairwise, *e.g.*, asymmetric C-atoms. The parities of their permutation descriptors can be used as their *parity descriptors*. A permutation containing an even number of transpositions[p] has a parity of $+1$, conversely, odd permutations have parities of -1. Since α-atom RASI (see Section 5.4.1) correspond in most casses to the CIP sequences, an assignment of $+1$ to an asymmetric C-atom is equivalent to an R-configuration, and -1 to the S, if the skeleton is indexed as shown in *25*.

6. The Simultaneous Representation of Molecular Constitution and Stereochemical Features and their Interconversions

An empirical formula indicates a set A of n atoms which is contained in a given molecule, or ensemble of molecules (EM). Thus an EM(A)

---

o) A regular PI proceeds without breaking and making of bonds, by bond angle deformations or internal rotations.

p) A transposition is the exchange of two adjacent elements of an ordered set. A cyclic permutation of n symbols contains $n-1$ transpositions.

is any compound or collection of chemical species that can be formed from A using each atom belonging to A exactly once. The constitutional formula of an EM(A) is the set of formulas of all molecules belonging to that EM(A).

An FIEM(A), the *family of isomeric ensembles of molecules* of A, is the set of all EM(A), Thus, an FIEM is described simply by a gross, empirical formula; an EM is described by a list of molecular species in terms of constitutional formulas. The constitutional chemistry of a set of atoms A can be described by the structure of its FIEM(A).

The chemical constitution of a molecule or an ensemble of molecules (EM) of n atoms is representable by a symmetric $n \times n$ BE-matrix and corresponds accordingly to a point P in $\mathbb{R}^{n(n+1)/2}$ an $n(n+1)/2$ dimensional Euclidean space, the Dugundji space of the FIEM(A). The "city block distance" of two points P_1 and P_2 is twice the number of electrons that are involved in the interconversion $EM_1 \rightleftharpoons EM_2$ of those EM that belong to the points P_1 and P_2. This chemical metric on the EM of an FIEM provides not only a formalism for constitutional chemistry, but also allows us to use the properties of Euclidean spaces in expressing the logical structure of the FIEM, and thus of constitutional chemistry [3e,32c].

Each point in a Dugundji space represents a chemical constitution for which there can exist a multiplicity of stereoisomers, corresponding to a set of CC-matrices associated with the BE-matrix of the constitution[q]. Owing to their interdependence, the constitutional and stereochemical features of chemical systems do not form a cartesian product, whereas the configurations and conformations can be represented as cartesian products.

This unified model of the logical structure of chemistry can be used as the basis of a unified representation of the essential features of chemical systems and their interconversions by chemical reactions. This requires a common indexing system as a reference. The atomic indexing of a constitution requires a different procedure than the indexing of the ligands of configurations by the CIP rules (see Section 2.3). This would correspond to combining different and independent nomenclature systems. A complete, separate treatment of the constitution and stereochemical features is, however, not possible, because the representation of a configuration or conformation must include a statement about the parts of the constitution that participate in the considered stereochemical feature.

The ASI and RASI provide the common reference system that is needed for the full use of the above model of the logical structure of chemistry for a computer oriented representation.

---

[q] The various types of relations between the stereoisomers belonging to a constitution are representable by graphs [32c].

6.1. Canonical Matrix Pairs and their Transformations

A universal, computer oriented representation of the relevant structural features of molecular systems and their interconversion by chemical reactions is possible with the pairwise combinations (E, C) of BE-matrices and CC-matrices. The transformation

$$(E, C) \xrightarrow{(Q,T)} (E', C')$$

of the starting material matrix pair (E, C) by the operator pair (Q, T) to the product matrix pair represents a chemical reaction, which is the transition from a given starting material point to a product point belonging to the Dugundji space and the associated stereoisomer spaces. Here Q acts upon E and T upon C.

$$QE = E'$$

$$TC = C'$$

Let E be a canonical BE-matrix (see Section 2.1) then the transformation $E \rightarrow E'$ involves two operations, the representation of the constitutional change by the addition of an R-matrix[3)] and a subsequent row/column permutation of E which restores the canonical form of the product BE-matrix[r)].

$$E + R = E$$

$$QE = E'$$

Normally, the stereochemical features of the reaction type and the starting materials determine the stereochemical features of the products. Accordingly one can derive the operator T from E and R, and thus the transformation $C \rightarrow C'$. Such a representation requires a correlation of the ligand indices of the stereochemical representation used and the indices of the atoms that enter the representation of the constitution.

A consistent assignment of indices to atoms of a constitution and ligands of configurations is possible on the basis of the ASI and the RASI of the ligand α-atom. There is no need for defining the ligands explicitly. This is particularly advantageous in the description of reacting systems, where the concurrent constitutional changes have the effect that the

[r)] The latter operation is a coordinate permutation of the Dugundji and stereoisomer spaces.

skeletons and ligands are different before and after the reaction. Although in reactions the atom ASI and RASI of configurations may differ between starting material and product, they are easy to correlate[3g].

6.2. Representation of the Stereochemical Aspect by Parity Vectors and van't Hoff's Concept

In the treatment of stereochemical aspects for many chemical problems such as synthetic design representation of tri- and tetracoordinate monocentric configurations by their parity descriptors suffices. The conformational aspect as well as higher coordinate and polycentric configurations can be neglected. Then, it is possible to reduce the CC-matrices to parity vectors P_n whose components $+1$, 0, -1 represent the configurational features.

Representation of molecular configuration by parity vectors relates directly to van't Hoff's concept of superposition of asymmetric C-atoms. The transformations

$$(E, P_n) \xrightarrow{(Q,S)} (E', P_n')$$

affords a simultaneous treatment of the constitutional and stereochemical aspects in a computer program for synthetic design on the basis of a mathematical model, without the use of detailed chemical information.[3g] The stereochemical changes in chemical reactions are particularly easy to describe by the algebra of parity vectors. All essential features, including the Woodward-Hoffmann [35] rules and related rules can be expressed in terms of 0 and ± 1.

In essence the parity vectors are based upon van't Hoff's concept of decomposing polycentric configurations into asymmetric carbon subunits, both treatments correspond in scope and limitations.

Acknowledgments. We wish to thank Ms. J. Blair, Ms. C. Gillespie, Prof. G. Snatzke, and Prof. J. Dugundji for helpful discussions, and we acknowledge the support of our work by Deutsche Forschungsgemeinschaft, Stiftung Volkswagenwerk and Fonds der Chemischen Industrie.

7. References

1) Harary, F.: Graph theory. Reading, Mass.: Addison-Wesley Co. 1972.
2) a) Spialter, L.: J. Am. Chem. Soc. *85*, 2012 (1963);
b) Spialter, L.: J. Chem. Doc. *4*, 261, 269 (1964).
3) a) Ugi, I., Gillespie, P.: Angew. Chem. *83*, 980 (1971); Angew. Chem. Intern. Ed. Engl. *10*, 914 (1971);

b) Ugi, I., Gillespie, P.: Angew. Chem. *83*, 982 (1971); Angew. Chem. Intern. Ed. Engl. *10*, 915 (1971);
c) Ugi, I.: Intra-Sci Chem. Rep. *5*, 229 (1971);
d) Ugi, I., Gillespie, P. D., Gillespie, C.: Trans. N. Y. Acad. Sci. *34*, 416 (1972);
e) Dugundji, J., Ugi, I.: Topics Curr. Chem. *39*, 19 (1973);
f) Blair, J., Gasteiger, J., Gillespie, C., Gillespie, P., Ugi, I.: In: Proceedings of the NATO/CNA ASI on computer representation and manipulation of chemical information (ed. W. T. Wipke, S. Heller, R. Feldmann, and E. Hyde). New York: John Wiley 1973;
g) Blair, J., Gasteiger, J., Gillespie, C., Gillespie, P. D., Ugi, I.: Tetrahedron, submitted (1973);
h) Blair, J., Gasteiger, J., Gillespie, C., Gillespie, P. D., Ugi, I.: In: Methodicum chimicum (ed. R. Gompper and H. Machleidt). Stuttgart: G. Thieme Verlag (in preparation).

4) a) Cahn, R. S., Ingold, C. K., Prelog, V.: Experientia *12*, 81 (1956);
b) Cahn, R. S., Ingold, C. K., Prelog, V.: Angew. Chem. *78*, 413 (1966); Angew. Chem. Intern. Ed. Engl. *5*, 385 (1966).

5) Morgan, H. L.: J. Chem. Doc. *5*, 107 (1965).

6) Blair, J.: in preparation.

7) Ugi, I., Marquarding, D., Klusacek, H., Gokel, G., Gillespie, P.: Angew. Chem. *82*, 741 (1970); Angew. Chem. Intern. Ed. Engl. *9*, 703 (1970).

8) a) Eliel, E. L.: Stereochemistry of carbon compounds. New York: McGraw-Hill 1962;
b) Hirschmann, H., Hanson, K. R.: J. Org. Chem. *36*, 3293 (1971).

9) Pólya, G.: Acta Math. *68*, 145 (1937).

10) a) Ruch, E.: Theoret. Chim. Acta *11*, 183 (1968);
b) Ruch, E.: Acc. Chem. Res. *5*, 49 (1972);
c) Ruch, E., Runge, W., Kresze, G.: Angew. Chem. *85*, 16 (1973); Angew. Chem. Intern. Ed. Engl. *12*, 25 (1973).

11) a) Lord Kelvin: Baltimore lectures, p. 436, 619. London: C. J. Clay and Sons, 1904;
b) McWeeny, R.: Symmetry. New York: Pergamon Press 1963.

12) Biot, J. B.: Bull. Soc. Philomath. Paris, *1815*, 190; *1816*, 125, Chem. Acad. Roy. Sci. Inst. France *2*, 41, 114 (1817).

13) Herschel, J. F. W.: Trans. Cambridge Phil. Soc. *1*, 43 (1821).

14) Pasteur, L.: Two lectures delivered before the Societe Chimique de Paris, Jan. 20 and Feb. 3, 1860.

15) Kekulé, A.: Liebigs Ann. Chem. *106*, 154 (1858).

16) van't Hoff, J. H.: Bull. Soc. Chim. France, [2] *23*, 295 (1875) (the original version appeared in Dutch in 1874).

17) Le Bel, J. A.: Bull. Soc. Chim. France [2], *22*, 337 (1874).

18) a) Gould, E. S.: Mechanismus und Struktur in der organischen Chemie. Weinheim/Bergstr.: Verlag Chemie 1962;
b) Roberts, J. D., Caserio, M. C.: Basic principles of organic chemistry. New York: W. A. Benjamin 1965;
c) Fieser, L. F., Fieser, M.: Organische Chemie. Weinheim/Bergstr.: Verlag Chemie 1965;
d) March, J.: Advanced organic chemistry: reactions, mechanisms, and structure. New York: McGraw-Hill 1968;
e) Hendrickson, J. B., Cram, D. J., Hammond, G. S.: Organic chemistry. New York: McGraw-Hill 1970.

19) van't Hoff, J. H.: The arrangement of atoms in space. London: Longmanns, Green 1898.

20) a) Mislow, K.: Bolstadt, R.: J. Am. Chem. Soc. *77*, 6712 (1955);
b) Mislow, K.: Science *120*, 232 (1954); Trans. N.Y. Acad. Sci *14*, 298 (1957);
c) Mislow, K.: Einführung in die Stereochemie. Weinheim/Bergstr. Verlag Chemie 1967.

21) Aschan, O.: Chem. Ber. *35*, 3389 (1902).

22) Prelog, V., Helmchen, G.: Helv. Chim. Acta *55*, 2581 (1972).

23) a) Klages, F.: Lehrbuch der organischen Chemie, Vol. II, p. 529. Berlin: Walter de Gruyter 1954;
b) Karrer, P.: Lehrbuch der organischen Chemie, p. 136. Leipzig: VEB Georg Thieme 1954;
c) Hückel, W.: Theoretische Grundlagen der organischen Chemie, Vol. I, p. 71. Leipzig: Akademische Verlagsgesellschaft Geest und Portig 1956;
d) Mann, F. G.: Progr. Stereochem. *2*, 196 (1958);
e) Noller, C. R.: Lehrbuch der organischen Chemie, p. 368. Berlin-Göttingen-Heidelberg: Springer 1960.

24) Tröger, J.: J. Prakt. Chem. [2] *36*, 225 (1887).

25) Prelog, V., Wieland, P.: Helv. Chim. Acta *27*, 1127 (1944).

26) Park, C. H., Simmons, H. E.: J. Am. Chem. Soc. *94*, 7184 (1972).

27) Marquarding, D., Ugi, I.: In: Methodicum chimicum (ed. R. Gompper and H. Machleidt). Stuttgart: G. Thieme Verlag (in preparation).

28) Fischer, E.: Chem. Ber. *24*, 2683 (1891).

29) Fresnel, A.: Ann. Chim. Phys. [2] *28*, 147 (1825).

30) a) Ugi, I.: Jahrb. 1964 Akad. Wiss., S. 21. Göttingen: Vandenhoek u. Rupprecht 1965;
b) Ugi, I.: Z. Naturforsch. *20b*, 405 (1965);
c) Ugi, I., Offermann, K., Herlinger, H.: Chimia (Aarau) *18*, 278 (1964);
d) Ruch, E., Ugi, I.: Topics Stereochem. *4*, 99 (1969).

31) Blackwood, J. E., Gladys, C. L., Loening, K. L., Petrarca, A. E., Rush, J. E.: J. Am. Chem. Soc. *90*, 509 (1968).

32) a) Gillespie, P., Hoffmann, P., Klusacek, H., Marquarding, D., Pfohl, S., Ramirez, F., Tsolis, E. A., Ugi, I.: Angew. Chem. *83*, 691 (1971); Angew. Chem. Intern. Ed. Engl. *10*, 687 (1971);
b) Ugi, I., Marquarding, D., Klusacek, H., Gillespie, P., Ramirez, F.: Acc. Chem. Res. *4*, 288 (1971);
c) Dugundji, J., Gillespie, P. D., Marquarding, D., Ugi, I.: In: The chemical application of graph theory (ed. A. Balaban). London: Academic Press (submitted 1971, in press);
d) This representation was recently also reproduced by Klemperer, W. G.: J. Am. Chem. Soc. *94*, 6940 (1972).

33) a) Longuet-Higgins, H. C.: Mol. Phys. *6*, 445 (1963);
b) Hougen, J. T.: Pure Appl. Chem. *11*, 481 (1965).

34) Ruch, E., Hässelbarth, W., Richter, B.: Theoret. Chim. Acta *19*, 288 (1970).

35) a) Evans, M. G., Polanyi, M.: Trans. Faraday Soc. *34*, 11 (1938). — Evans, M. G., Warthurst, E.: Trans. Faraday Soc. *34*, 614 (1938). — Evans, M. G.: Trans. Faraday Soc. *35*, 824 (1939);
b) Schuler, K. E.: J. Chem. Phys. *21*, 624 (1953);
c) Dewar, M. J. S.: Tetrahedron Suppl. to *8*, 75 (1966). — Dewar, M. J. S.: Angew. Chem. *83*, 859; Angew. Chem. Intern. Ed. Engl. *10*, 761 (1971);
d) Woodward, R. B., Hoffmann, R.: Angew. Chem. *81*, 797 (1969); Angew. Chem. Intern. Ed. Engl. *8*, 781 (1969);

e) van der Hart, W. J., Mulder, J. J. C., Oosterhoff, L. J.: J. Am. Soc. *94*, 5724 (1972). — Mulder, J. J. C., Oosterhoff, L. J.: Chem. Commun. *1970*, 305, 307;
f) Fukui, K.: Acc. Chem. Res. *4*, 57 (1971);
g) Anh, N. T.: Die Woodward-Hoffmann-Regeln und ihre Anwendung. Weinheim/Bergstr.: Verlag Chemie 1972.
h) Gilchrist, T. L., Storr, R. C.: Organic reactions and orbital symmetry. London: Cambridge University Press 1972.

Received October 4, 1973

Stereospecificity in Biology

Prof. Dr. Birgit Vennesland

Forschungsstelle Vennesland der Max-Planck-Gesellschaft, Berlin-Dahlem

Contents

"Thought — to call it by a prouder name than it deserved — had let its line down into the stream. It swayed, minute after minute, hither and thither among the reflections and the weeds, letting the water lift it and sink it, until — you know the little tug — the sudden conglomeration of an idea at the end of one's line: and then the cautious hauling of it in, and the careful laying of it out? Alas, laid on the grass how small, how insignificant this thought of mine looked; the sort of fish that a good fisherman puts back into the water so that it may grow fatter and be one day worth cooking and eating. I will not trouble you with that thought now, though if you look carefully you may find it for yourselves in the course of what I am going to say.

But however small it was, it had, nevertheless, the mysterious property of its kind — put back into the mind, it became at once very exciting, and important; and as it darted and sank, and flashed hither and thither, set up such a wash and tumult of ideas that it was impossible to sit still."

Virginia Woolf in

A Room of One's Own

Introduction

Virtually all biological reactions are stereospecific. This generalization applies not only to the enzyme-catalyzed reactions of intermediary metabolism, but also to the processes of nucleic acid synthesis and to the process of translation, in which the amino acids are linked in specific sequence to form the peptide chains of the enzymes. This review will be restricted mainly to some of the more elementary aspects of the stereospecificity of enzyme reactions, particularly to those features of chirality which have been worked out with the help of isotopes.

The availability of isotopes has made it possible to complete the descriptions of the steric course of most of the individual reactions of carbohydrate metabolism and steroid metabolism and many of the reactions of fat and amino acid metabolism. The subject has been covered from various angles in several chapters of the third edition of the Enzymes, particularly in the one by Popjack [1], in a comprehensive treatise [2,3], and in numerous recent reviews [4–12]. The wealth of available detail defies any attempt to be complete. I will try, rather, to describe trends in current experimentation, and to fit these trends into historical perspective. In so doing, I will select examples rather arbitrarily, entirely out of my own interests, and I beg the reader's indulgence for this bias.

After the above statement of intentions was prepared, I realized it wasn't very honest, in the sense that it wasn't quite going to prepare the reader for what was coming. It was then that I thought of the introduction by Virginia Woolf.

Some History

One could say that Pasteur discovered chirality and that van't Hoff identified it in terms of structural organic chemistry. It had the compelling attractiveness of mystery about it, and when the enzymes were identified as the catalysts which cells employed to make optically pure compounds, the question "How in the world do they do it?" (Westheimer, unpublished) was a question that fired the imagination. This question has, in principle, now been answered by the X-ray crystallographers and the chemists who are elucidating the complete three-dimensional structure of the proteins. The lecture delivered by Perutz in 1968 at the 5th FEBS Meeting in Prague [13] describes with simplicity and logical satisfaction how the stereospecificity and high catalytic power of enzymes can both be accounted for by their three-dimensional structure. His description of the stereochemistry of the cooperative effects in haemoglobin [14] further illustrates the power of the method for analyzing the nature of conformational changes in protein. This reviewer can only watch in admiration as the anatomy of enzyme after enzyme is described with three-dimensional models and stereodiagrams [6,8,9,15] [a]. One sees how the peptide chains of the protein molecules are folded and twisted in such a way as to form reaction centers where substrates are oriented very precisely toward each other and toward activating groups. The stereospecificity is an aspect of the catalytic mechanism [16]. Further speculation in this area is unnecessary. One need only direct the questions to the X-ray crystallographers, and the mystery had moved to another level. Perhaps, for example it is possible to conserve the energy of a redox reaction (as in the respiratory chain) in the form of a conformational change of the protein [8].

The Rediscovery of the Chiral Center

But back to isotopes and stereospecificity. Popjack [1] has described in admirable detail the sequence of events which began in 1940 with the application of ^{11}C and ^{13}C to the study of carbohydrate metabolism, and culminated with Ogston's discovery, in 1948, that isotopes could indeed give very special information about the stereospecificity of enzyme reactions. In this narrative, Popjack notes that it seems strange, in retrospect, that so much time elapsed before Ogston's discovery. This is a polite way of saying that Ogston's discovery should have been self-evident in 1940. I intend no discredit to Ogston, to whom I am very deeply indepted. Since I was a close spectator, and almost a participant in the events in question, I'd like to try to explain why it took so long. It is, I think, an interesting example of the importance of language for clear thinking, particularly of the paramount importance of clear and unambiguous definitions.

Isotopes were not available in van't Hoff's day. My student generation was taught that an asymmetric carbon atom was a carbon atom attached to 4 chemically different groups. When isotopes of carbon, ^{11}C and ^{13}C, were first applied as tracers to study carbohydrate metabolism, the entire emphasis was on the chemical similarity of ^{11}C and ^{13}C to the more abundant isotope ^{12}C. Thus, it was of pressing interest to determine whether CO_2 participated in the oxidation of carbohydrate in animal tissues, a conclusion strongly suggested by the demonstration in Krebs' laboratory, that pyruvate and oxalacetate behaved alike in pigeon liver, and by Wood and Werkman's earlier demonstration that some he-

[a] Volume 5 of the Atlas of protein sequence and structure was published in 1972.

terotrophic bacteria could utilize CO_2. The usefulness of isotopes for answering such a question lay in the fact that ^{11}C from $^{11}CO_2$ would behave like ^{12}C from $^{12}CO_2$, but that ^{11}C could be detected in isolatable intermediates, even if there was a net loss of CO_2 in the overall reaction.

And so it happened that the detection of isotopic carbon from $^{11}CO_2$ in the β carboxyl group of α-ketoglutarate was celebrated as a discovery of CO_2 fixation in animal tissues, and was taken as evidence that there wasn't any citric acid in the citric acid cycle. A citric acid precursor of α-ketoglutarate would have to be labelled as in compound *1*

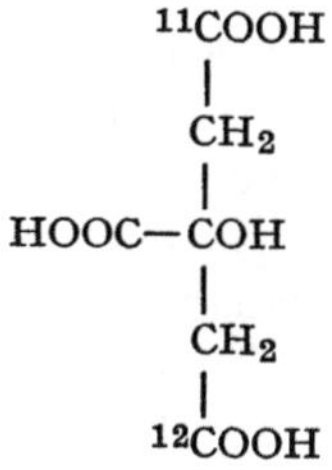

1

and compound *1* was obviously a symmetrical molecule, because there wasn't any chemical difference between ^{11}C and ^{12}C. Even an enzyme couldn't tell any chemical difference between ^{11}C and ^{12}C. That was the principle on which tracer methodology was based. See what I mean! As evidence that the problem was one of language, I quote from Ogston's paper:[16a)]

"...It is possible that an asymmetric enzyme which attacks a symmetrical compound can distinguish between its identical groups".

Ogston referred to

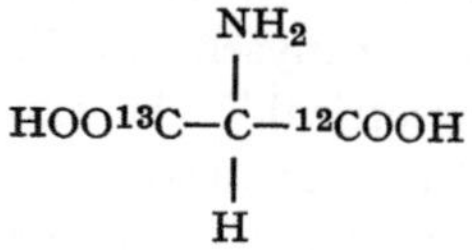

as a symmetrical compound, though he was the first to appreciate that it was in fact not symmetrical. He had to call it symmetrical, though, because it was symmetrical by definition, and we all have to follow the rules and use the same definitions. In an analogous but rather different situation, Wood and Werkman's clear demonstration that a heterotrophic bacterium could utilize CO_2 (done without isotopes) was regarded with disbelief for years, though their experiments were really very good and convincing, because heterotrophs, by definition, were organisms which required carbon in organic form, and couldn't utilize CO_2.

When I first read Ogston's paper, I thought I understood it, but in actual fact, I did not. I didn't at the time find aminomalonic acid interesting[b]. I was compelled to careful reflection, however by a combination of circumstances. A student in the department had obtained some chemical evidence that citrate was the first tricarboxylic acid product formed from oxalacetate and pyruvate. This was contrary to all expectations; an apparent complete contradiction of other, better experiments, done in the same department. I was assigned to find out what was wrong with the student's experiments, and I could not find anything wrong. The student stuck to his guns as a brave soldier should, and insisted his experiments were telling him citrate, and instead of my convincing him he was wrong, he was succeeding in convincing me that he was right, which is not a proper relationship between a professor and a student. You can see I was caught on the horns of a painful dilemma. In a mood of desperation, I sat down with Ogston's paper and took some stick and ball models to assist my thinking. Ogston's paper didn't help, though I read it several times attentively. I thought he was implying a substrate might stay stuck to the enzyme, and that did not help one bit. While I was staring bleakly at the models, they suddenly told me the answer, but not in words, just in a picture. What happened in that moment of sudden insight seemed to be that I was suddenly graced with the ability to see the asymmetric carbon atom as van't Hoff had originally seen it. The feeling I experienced was a curious combination of exhilarating, sweet relief (because there was no contradiction between the two sets of experiments) and total dismay: "Oh you idiot, why haven't you realized that before". There were quite a few of us who had experiences similar to mine after the appearance of Ogston's paper. Even though I felt I had grasped the point quite well I found that when I tried to transmit my comprehensions to others it was like the tower of Babel. I distinctly remember hearing myself say "Of course I know it is symmetrical, but it isn't symmetrical." The explanation that worked best with students was to say: "Don't you see — *look* at the model — *look* at it! The enzyme doesn't have to tell ^{11}C and ^{12}C apart chemically. It can feel they are in different positions. It's the same thing when a particular enzyme reduces pyruvate only to L-lactate — never D. The enzyme can tell that pyruvate has two sides, a left and a right." By a curious coinci-

b) Rat liver aminomalonate decarboxylase has recently been identified with cytoplasmic serine hydroxymethylase and allothreonine aldolase [17]. Most references in this narrative have been omitted, since they can all be found in Popjack's review [1].

dence, Harvey Fischer had just asked Frank Westheimer for a problem we could collaborate on, and Westheimer then asked me if it was practical to use deuterium to see if the alcohol dehydrogenase reaction proceeded with direct hydrogen transfer from alcohol to NAD. The stereospecificity of hydrogen transfer then fell on us like gentle rain from heaven [17a,17b]. But we never talked about prochiral centers. We only said "*Look, look,* there is another one."

Actually, anyone considering models, as Ogston obviously did, could easily grasp that compound *1* has a mirror image on which it is not superimposable. The reason this had not been grasped was that the wrong answer seemed so totally and obviously right, that no one had felt it was necessary to consult models.

And it took longer to devise an adequate nomenclature [18–22] for talking about the stereochemical insight provided by isotopes, than it took for biochemists to discover they had forgotten their van't Hoff. The invention of language is man's greatest and maybe his most difficult invention. We still have a long way to go in developing communication.

I will try to adhere to the IUPAC/IUB rules for stereochemical nomenclature [22], though I am inherently suspicious of all rules and definitions. The terms chirality and prochirality are so obviously fit that they have been generally adopted and explained in many places, and the R/S nomenclature is so essential for describing chiral centers involving isotopes, that I'm grateful for that too, though I'd be still more grateful if someone could dream up a way of making the assignment of "priority" easier. My difficulty is that I wish chemistry were simple.

Stereochemistry in Biology

The study of the mechanism of vision in vertebrates [23,24] has progressed to the point where the first consequence of photon absorption has been described as an activation of the isomerization of the 11-*cis* retinal chromophore of rhodopsin to all-*trans*. That triggers a complex sequence of reactions leading to the mysterious inside of the brain. Brrr, I had better get back — it looks dark in there. But the brain can generate sensations of light. Maybe, one day, we will be able to see enough to understand, but we'll go back just the same to a safer subject.

One could plunge into the steric problems posed by the mechanism of protein synthesis on the ribosome [25,26], or consider the steric fit of the hormone insulin to its acceptor in the cell membrane [27]. Or one could delve into the beautiful intricacy of terpenoid, squalene and steroid metabolism, or get lost in double bond formation, or in the steric problems posed by the branched chain fatty acids and their derivatives [28–34].

There are some very interesting questions of stereospecificity posed by the structure and mode of operation of multienzyme complexes. Reed and Cox [35] have summarized available information on the pyruvate and α-ketoglutarate dehydrogenase complexes, and the fatty acid synthetase. The mechanism of synthesis of the peptide antibiotics likewise presents interesting stereochemical problems [36].

The use of stereochemical concepts for the description of the association of subunits in oligomeric and polymeric proteins has been discussed by Hanson [7], who has provided a clarifying summary of terminology, as well as some interesting examples.

Stereospecificity of Enzyme-Catalyzed Reactions

With isotopes it has been possible to show that all enzyme-catalyzed reactions are stereospecific. Before the availability of isotopes, there was no way of testing this generalization. Of course there are some apparent exceptions to prove the rule. Bently has listed a considerable number ([2], Table XIII, Chapter 6). The most interesting one to me seems to be luciferase, but that is an exception that isn't an exception. Thus, the enzyme luciferase acts on its substrate luciferin (*2*), in the presence of ATP and O_2, to oxidize the luciferin to oxyluciferin (*3*). The reaction consists of an initial activation of the substrate by ATP to give luciferyl adenylate, after which the oxidation takes place. When the natural enantiomer (synthesized from D-cysteine) is activated and oxidized, light is emitted. The other enantiomer is also acted on by the enzyme, and is converted to the adenylate, but oxyluciferin is not formed, and there is no bioluminescence [37,38,38a].

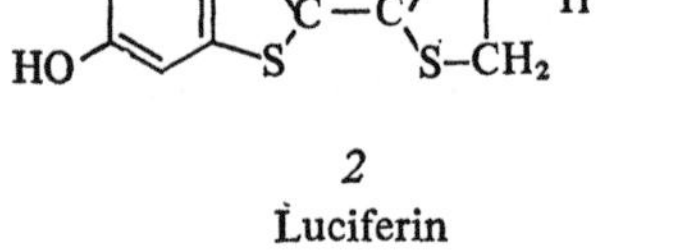

2
Luciferin

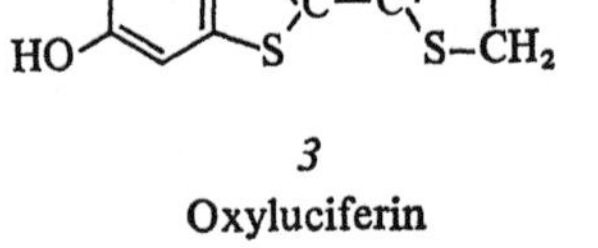

3
Oxyluciferin

Citrate Synthase

The stereochemistry of enzymic citrate synthesis and cleavage has been described in detail [39]. In microorganisms, it has been found that the stereospecificity of citrate synthase from different sources is different.

O'Brien and Stern [40] have shown that the enzyme from *Clostridium kluyveri* has R-specificity, but that this can be changed reversibly to S-specificity by subjecting the enzyme to oxidizing conditions or to the —SH binding reagent, pCMB. The stereospecificity of the reaction is apparently determined by one or more —SH groups in the molecule. It appears, as some of the exceptions to the general rules of complete stereospecificity are challenged, that it will be possible to explain them away. Thus, an enzyme causing a racemization may proceed by sterically directed reactions [41]. Furthermore, an enzyme may direct the reduction of pyruvate by $NaBH_4$ to give a preponderance of one enantiomer of lactate, even though the reduction of pyruvate is not the natural function of the enzyme. This has been used to obtain information about the stereochemistry of binding of pyruvate to oxaloacetate decarboxylase [42] and to pyruvate kinase [43].

UDP Galactose-4-Epimerase

Interesting stereochemical studies have been done with UDP-galactose-4-epimerase [44–49], which catalyzes the interconversion of UDP-galactose and UDP-glucose. The yeast and *E. coli* epimerases have been purified and extensively investigated. They both contain one mole of tightly bound NAD per mole of enzyme. It was originally proposed by Maxwell [44] that the enzyme functions by transferring H from C—4 of the carbohydrate to NAD, and then back from the NADH to the carbonyl group at C—4 of the carbohydrate. This type of mechanism has been well substantiated [48], and it has been shown that the H transfer is stereospecific for the 4 pro S or 4 B position of the nicotinamide ring of the NADH. The problem then was to explain the lack of stereospecificity of the enzyme for the C—4 position of the sugar. It has been proposed that the carbohydrate can bind in two modes or positions [47,49] an idea supported by the fact that the enzyme can transfer H from C—1 (as well as C—4) of certain sugars, to give aldonic acids. Such reactions lead to inactivation of the enzyme [45,46] and it has been suggested that they have a regulatory function.

Rose [11] has distinguished between substrate specificity (which may not seem stereospecific) and reaction stereospecificity (which is always present in enzyme reactions) in a thought-provoking essay, in which he also describes the usefulness of stereospecificity generally in studying enzyme mechanisms. He has done this so much better than I could, that I will not try to add to this subject. He has also pointed out how stereospecificity may be useful in the study of evolution. This is a subject I would like to amplify. The less we know, the more we say. Nevertheless, the Journal of Molecular Evolution, which began publication in 1971, reflects the current interest in this subject.

Stereospecificity and Evolution

The modern view of the origin of life seems to be that, given the right conditions, it was inevitable rather than improbable [50–54]. We may assume that systems of life analogous to our own exist elsewhere in the universe, wherever there are favorable conditions, and of all these systems, half would have the same configuration as our own. The others would be mirror images. The two systems could not develop together in the same place because they would compete with each other rather than support each other, and one or the other would win out [55]. To form helical structures, which reproduce best, the members of the coiled chain must have the same configuration, and chains of nucleotides of the same configuration would give chains of amino acids of the same configuration. Efforts are underway to explain the origin of the genetic code [52–54,57] and it is probably going to be possible to explain why the amino acids of the proteins must have the L-configuration if the pentoses of the nucleic acids have the D-configuration.

One of the groups of theories about the origin of the genetic code states that the code has to be the way it is, and is therefore universal, for "stereochemical" reasons. In other words, phenylalanine, f. ex. *must* be represented by the triplets UUU and UUC because phenylalanine is somehow stereochemically related to these two codons [52,53,56,57]. This seems likely, since steric fit is an essential property of the processes of replication, transcription and translation. That doesn't mean that one has conclusive evidence for such a statement. It only means that the theoreticians are groping in such a direction.

We move up the ladder of life with the assumption that the primitive ancestral cell, living in an organic soup, had relatively few enzymes, and that these few enzymes have evolved, by gene duplication and mutation, into other enzymes, so that all enzymes may eventually be classified into a relatively few large families. Amino acid sequence analysis and X-ray crystallography should one day make it possible to trace many such enzyme family trees. Substantial success has been achieved with some of the proteolytic enzymes [58,59], but the pioneering work of Margoliash and Dickerson and their associates on cytochrome *c* probably gives the clearest foretaste of what can eventually be done [8,60–63].

Though it is a protein, cytochrome *c* is, strictly speaking, not an enzyme, but a substrate for other enzymes of the respiratory chain. Margoliash, Smith, Boulter and others have determined the complete amino acid sequences of cytochrome *c* for over 35 species. The structure of the molecule has been determined to a resolution of 2,8 A [61,62]. About 35 residues (out of about 104) are highly conserved. Residues 70—80 are totally invariant, and are regarded as a key component of one of the redox mechanisms of the protein, being closely associated with the heme prosthetic group. A large part of the surface of the molecule is structured and conserved through evolution, presumably because cytochrome *c* has a large area of contact with its reductase on the one side, and its oxidase on the other [61]. It has also been demonstrated, with purified sitespecific antibodies, that there are separate oxidase and reductase reaction sites on the surface of cytochrome *c* involved in its interaction with the oxidase and reductase segments of the respiratory chain [63].

The stereospecificity of an enzyme-catalyzed reaction may provide helpful clues to possible genetic relationships of different enzymes, because the stereospecificity would be expected to be conserved to a large extent during the evolutionary changes of the protein. Let us explore the possibility of using stereospecificity as a guide to trace enzyme family relations, by considering the pyridine nucleotide dehydrogenases.

Stereospecificity of the Pyridine Nucleotide Dehydrogenases

The pyridine nucleotide dehydrogenase reactions fall into two large groups: those which transfer hydrogen from or to the A (or pro R) side of the 4 position of the nicotinamide ring, and those which use the B (or pro S) side [1,64,65].

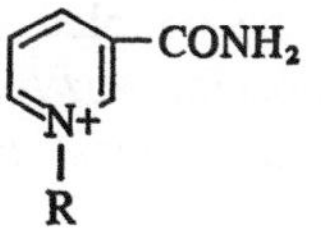

4
Oxidized

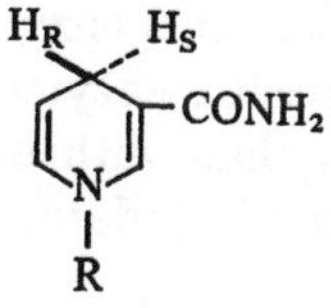

5
Reduced

Nicotinamide ring of pyridine nucleotides

We may note, parenthetically, that a somewhat similar possibility exists for the enzyme group in which the cofactor is pyridoxal phosphate or pyridoxamine phosphate. The stereochemistry of these reactions has been reviewed by Dunathan [5], who points out that during the enzyme reactions, there occurs a stereospecific protonation of the pyridoxamine methylene group. It might seem possible, in principle, that there would be a group of enzymes stereospecific for one of the hydrogen positions, and another group stereospecific for the other hydrogen position (see *6*). All of the five pyridoxal phosphate enzymes so far examined, however, show the same absolute stereospecificity of cofactor protonation, adding

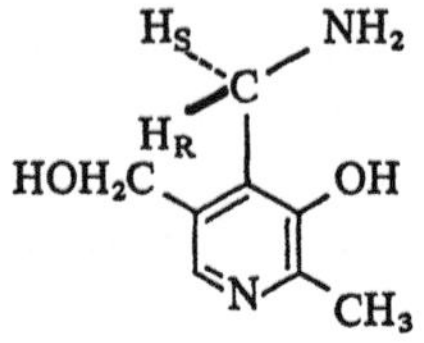

6
Pyridoxamine

the proton from the si face of C′4, as in *6*, so that the labile proton occupies the *pro*-S position of the pyridoxamine methylene group [66].

The 1969 Konstanz Symposium on Pyridine Nucleotide-Dependent Dehydrogenases [67] gives a good, overall picture of the kind of detailed information that is available about this large group of enzymes. At the time of that symposium, the structural studies of glyceraldehyde 3-phosphate dehydrogenase, a B-specific enzyme, and lactic dehydrogenase, an A-specific enzyme, were already well advanced [68,69]. Further progress is described in the Cold Spring Harbor Symposium [9].

A sufficient number of different dehydrogenases have now been examined to permit the generalization that there seem to be approximately equal numbers of A enzymes and B enzymes. An older suggestion of mine, that A-specific enzymes could be coupled more readily with B-specific enzymes than with other A-specific enzymes, has been tried and found wanting [70]. Indeed, in the cases investigated, the opposite seems to be true.

Do these A-specific enzymes and B-specific enzymes constitute two large families? This is a question that can not yet be answered, but one searches for family features which may characterize the two groups.

At this stage in the discussion, if this were a scholarly paper, we would produce a table listing all dehydrogenases found to be A-specific, and another table listing all dehydrogenases found to be B-specific. One can construct such tables by adding to Bentley and Popjack's tabulations [1,3], the extra information accumulated since these Chapters were written.

Hydroxymethylglutaryl CoA Reductase

The recent work on the stereospecificity of hydroxymethylglutaryl-CoA reductase [71–76] is particularly interesting. The reaction catalyzed by this enzyme is an important early step in the synthesis of terpenoids and steroids. The yeast enzyme and the liver enzyme both have the same stereospecificities. The overall reaction catalyzed is the reduction of hydroxymethylglutaryl CoA to mevalonic acid, as shown in scheme *1*. Two molecules of NADPH are used to reduce the Co-A-bound carboxyl

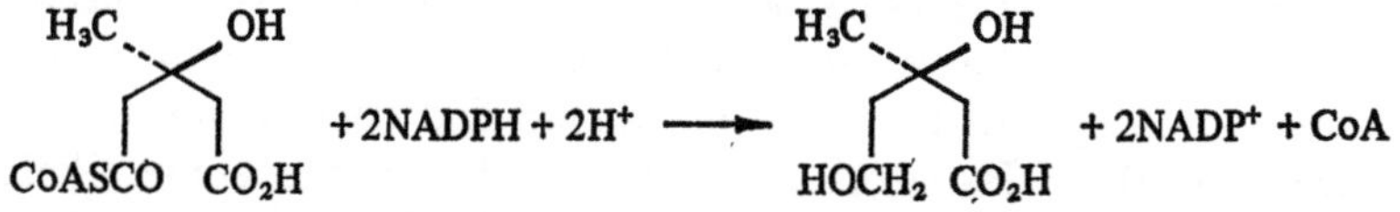

Scheme 1. Reaction catalyzed by hydroxymethylglutaryl CoA reductase

group to an alcohol. The reaction has been shown to occur in two steps [73,75]. The intermediate, mevaldic acid (in the form of a thiohemiacetal), is also a substrate for the enzyme. Both reduction reactions involve the transfer of H from the A (or pro R) side of the nicotinamide ring of NADPH to C–5 of the substrate. The second hydrogen transfer occurs stereospecifically to the pro S position at C–5 (see *7*). The liver enzyme and the yeast enzyme are assumed to be closely related because of these similar properties [74].

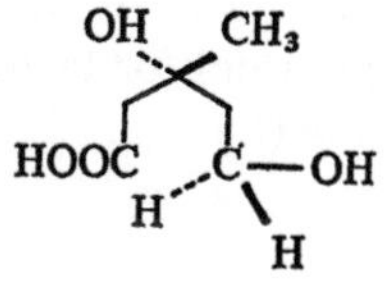

7

This is another example of the generalization that enzyme reactions of the same type have the same stereospecificity for the pyridine nucleotide, no matter what their cellular origin. There is another mevaldic reductase in the cytosol of liver, which catalyzes the reduction of mevaldic acid to mevalonic acid by NAD or NADP, and this enzyme also has A (or pro R) stereospecificity for the pyridine nucleotide [1] but the hydrogen transfer occurs to the pro R position on C–5. This latter enzyme can use either 3R or 3S mevaldic acid; that is, it is indifferent to the configuration of the substrate at C—3; whereas hydroxymethylglutaryl CoA reductase only operates on substrate with the 3S configuration in hydroxymethylglutaryl CoA or the 3R position in mevaldic acid. Oh you sequence rulers! These are really equivalent positions [1,73].

The true biological function of liver mevaldic reductase is not clear. It is not thought to be involved in cholesterol synthesis, and because of the difference in its stereospecificity for the substrate, it is thought to be only a distant relative of the hydroxymethylglutaryl CoA reductases. But all of these enzymes have the same A stereospecificity for the pyridine nucleotide.

There is some interesting new methodology that has been worked out in the course of these studies [73,76]. In general, stereospecificity relations of this sort are examined with the help of other enzyme reactions of known stereospecificity. When mevalonic acid (lactone form, *8* is fed to *Claviceps*, the 5-H_S atom is incorporated into the ergot alkaloids agroclavin *9a* and elymoclavin *9b*, whereas the 5H_R atom is lost.

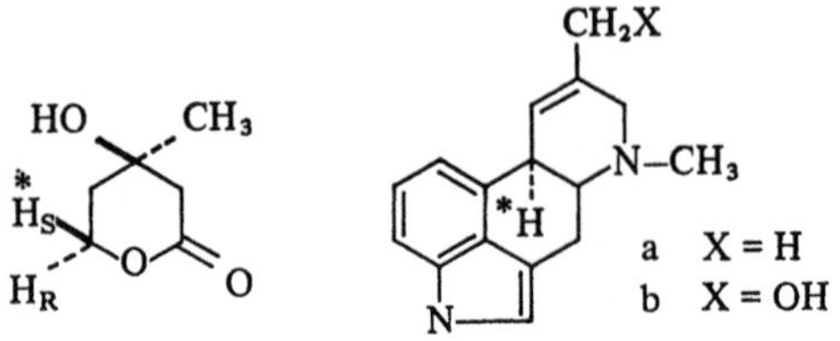

8 *9*

In the "classical" procedures [1], the 5-T or D-labeled mevalonate is converted enzymatically to farnesol, which is then oxidized to farnesal by liver alcohol dehydrogenase. This enzyme transfers the pro-R hydrogen of C—1 of ethanol or geraniol (or farnesol) to the 4 pro R position of the nicotinamide ring of NAD.

Alcohol Oxidation

The liver alcohol dehydrogenase mentioned in the preceding section has the same pro-R stereospecificity for NAD and ethanol as yeast alcohol dehydrogenase. Furthermore, the oxidation of ethanol by a microsomal oxidizing system, or by catalase and H_2O_2, likewise proceeds with pro-R stereospecificity for the ethanol [77]. The catalase-H_2O_2 system is so very different, however, from the pyridine nucleotide dehydrogenase, that one wonders whether the similarity in stereospecificity for ethanol is fortuitous.

Steroid Dehydrogenases

Gibb and Jeffery have made a detailed study of the stereospecificity of the reduction by NADH of 3-oxo steroids, catalyzed by cortisone reductase from *Streptomyces spp.*[78]. This enzyme had previously been shown to catalyze the transfer of the 20 α-hydrogen of 20 β-hydroxy-pregn-4en-3-one to the 4B-position of NAD^+ [79]. The enzyme has been highly purified, and the 20 α-dehydrogenase activity appears to be closely related to the 3-hydroxysteroid dehydrogenase activity [80,81]. It is unusual in not showing any absolute requirement at C—3 of the substrate for 3α-or 3ß- or for 3-axial or 3 equatorial, 3 R or 3 S [82,83]. Nevertheless, the enzyme binds NADH in such a way that only the 4 B or 4 pro R hydrogen is used in all the various possible reactions [78].

The stereospecificity of hydrogen transfer for estradiol-17α and estradiol-17β dehydrogenases has been examined by George *et al.*[84]. These enzymes are both present in chicken liver, and have substrates which differ only in the chirality of their substituents at C—17. Both of these enzymes were shown to use the 4-pro-S or 4B proton of the NADPH. Since the steroid is a bulky substrate, the authors argue that the steric fit between pyridine nucleotide and steroid cannot be as important as the role played by the enzyme in directing the fit. This paper contains an interesting summary of other recent work on the stereospecificity of pyridine nucleotide dependent-steroid dehydrogenases.

Transhydrogenases

The transhydrogenases present some very interesting stereochemical problems. These enzymes catalyze hydrogen transfer between two pyridine nucleotide molecules. Middleditch and Chung [85] have shown that the transhydrogenase from *Azotobacter vinelandii*, which catalyzes all the four possible reactions, NADH-NAD^+, NADPH-$NADP^+$, NADPH-NAD^+ and NADH-$NADP^+$, always uses only the 4 B or 4 pro S position of the reduced coenzyme. This is similar to the stereospecificity of the analogous enzyme from *Pseudomonas*, and different from the mammalian enzyme, which in contrast, has B or 4 S specificity for NADPH and A or 4 R stereospecificity for NADH. The transhydrogenases constitute an exception to the generalization that enzymes which catalyze the same reaction have the same stereospecificity. In the case of such an apparent exception, however, different mechanisms have been surmized: for example, one binding site only, for both substrates, when the stereospecificity for both substrates is the same, and two or more binding sites when it is different.

β-Oxidative Decarboxylases

Rose [11] has made an interesting survey of those enzyme reactions which can be characterized as electrophylic substitutions, and for which the occurrence of proton exchange, or the stereochemistry of proton exchange has been determined. In his Table 1, he lists these reaction in five groups (amino acid decarboxylases, aldolases, biotin-dependent carboxylases, ß-oxidative decarboxylases, and acetyl-CoA-like condensations). He points out that the ß-oxidative decarboxylase group is the only one which fails to show a common stereochemical course within the group. He has not included the available information about the stereospecificity for the pyridine nucleotide (the ß-oxidative decarboxylases are pyridine nucleotide requiring enzymes). In Table 1, I have added this information, which suggests that the ß-oxidative decarboxylases should be broken down into two groups of different origins: (a) those showing A stereospecificity for pyridine nucleotide and retention of substrate configuration, and (b) those showing B stereospecificity for pyridine nucleotide and inversion of substrate configuration. It is relevant also that the enzymes of the A group have a metal ion requirement whereas those of the B group do not [87].

Table 1. Stereochemistry of β-oxidative decarboxylases (references may be found in Rose [11]) and in Popjack [1])

	Stereospecificity for pyridine nucleotide	Proton replacement on substrate
Isocitrate (NAD-specific)	A	Retention
Isocitrate (NADP-specific)	A	Retention
6-phosphogluconate	B	Inversion
Malate	A	Retention
UDP glucuronate [1])	?	Inversion

[1]) This reaction involves transfer of hydrogen to protein-bound NAD with subsequent return of the hydrogen to the final substrate product, so that there is no net oxidation or reduction of NAD in the overall reaction [86]. The stereospecificity for the pyridine nucleotide would thus be difficult to determine. It has been done, however, in a similar situation, for UDP galactose-4-epimerase, which contains bound DPN which is reduced on the B side by the substrate (see text, p. 47). I once ventured a tentative generalization that all pyridine nucleotide dehydrogenase that acted on carbohydrates have B stereospecificity for the pyridine nucleotide. The ink was hardly dry on my paper before the first exception was noted. Now (we never learn), I want to try again. All pyridine nucleotide dehydrogenases which act on phosphorylated carbohydrate derivatives have B stereospecificity.

Generalizations Regarding Stereospecificity for Pyridine Nucleotides

Bentley [3] and Popjack [1] make three generalizations about the stereospecificity of hydrogen transfer to pyridine nucleotides.

1. The same enzyme reaction has the same stereospecificity, regardless of source. (This was obvious rather early in the game. Or should I say, if the stereospecificity were different, it would not be the same enzyme reaction.)
2. When an enzyme reacts with both pyridine nucleotides, NAD and NADP, it has the same stereospecificity for each of them. We can probably add to NAD and NADP, the analogues of either [88].
3. When an enzyme reacts with a range of substrates, the stereospecificity is the same with each substrate.

To me, these generalizations only seem to confirm what we know about enzyme active sites. It would have been much more surprising if the above facts had come out the other way around.

In searching for generalizations, I want something with a deeper meaning. I've tried to make a new one (Table 1). I like very much the interesting new generalization which has been proposed by Davies *et al.*[89]. They suggest that when a metabolic sequence involves consecutive nicotinamide-adenine dinucleotide-dependent reactions, the dehydrogenases have the same stereospecificity: *e.g.* glucose 6-phosphate dehydrogenase and 6-phosphogluconate dehydrogenase; aspartic β-semialdehyde dehydrogenase and homoserine dehydrogenase; nitrate reductase, nitrite reductase [c] and hydroxylamine reductase. All of the seven enzymes in these three sets are B-specific. Davies *et al.* note that the data support current theories on the evolution of enzymes, that the second dehydrogenase of the sequence is derived from the first, and thus likely to preserve the stereospecificity.

It should be noted, in this connection, that there are pyridine nucleotide dehydrogenases which catalyze redox reactions which must occur in two steps. Hydroxymethylglutaryl CoA reductase (discussed on p. 51) is one example. Another is uridine diphosphate-glucose dehydrogenase, which catalyzes the oxidation of the C—6 of the glucose (*i.e.*, a primary alcohol) to a carboxyl group. In both cases, there are two molecules of pyridine nucleotide required, and the overall reactions are essentially irreversible. The former enzyme, with A stereospecificity for the pyridine nucleotide, catalyzes the reduction of an acyl-CoA group

[c] Davies *et al.* used nitrite reductase of yeast. In leaves, nitrite reductase uses ferredoxin as a reductant, and is not a pyridine nucleotide dehydrogenase).

to an alcohol, and the latter enzyme with B stereospecificity [90] for the pyridine nucleotide catalyzes the oxidation of an alcohol to an acid. In both cases, the two consecutive steps of the reaction have the same stereospecificity for the pyridine nucleotide.

Biellman *et al.*[91] have reported that octopine dehydrogenase transfers hydrogen from the B side of the nicotinamide ring of NADH, or of the 3-cyano analogue of NADH. This enzyme catalyzes the oxidation of D octopine to L arginine and pyruvate, as shown in Scheme 2. Biellman *et al.* point out that the reaction catalyzed by octopine dehydrogenase is analogous to that catalyzed by glutamic dehydrogenase, which is also B-specific.

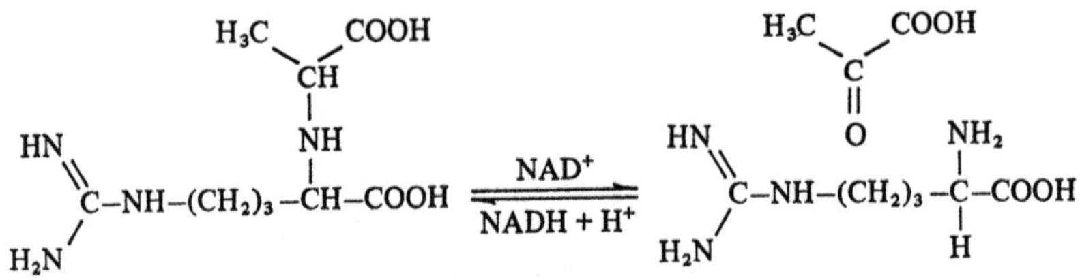

Scheme 2

In an earlier spectrophotometric study of this enzyme, a red shift of the reduced nicotinamide absorbance had been noted in the difference spectrum of the binding of reduced coenzyme to the purified protein. Fisher *et al.*[92] had pointed out that this is characteristic of most B-stereospecific dehydrogenases, so Biellman *et al.* have made a successful prediction for Fisher. Fisher's suggestion that the supernatant and mitochondrial forms of malate dehydrogenase have different stereospecificities for NAD^+ has not been substantiated, however [89].

Liver cells contain two different but closely related enzymes: glycerol phosphate dehydrogenase which is specific for NAD, and acylglycerol phosphate dehydrogenase, which is NADP specific. Both enzymes have B stereospecificity for the pyridine nucleotide [93]. They apparently have different metabolic functions.

The generalization that the same dehydrogenase has the same stereospecificity, no matter what the source of the enzyme, has been tested now particularly well for malic and lactic dehydrogenases. In fact, one can venture a guess, that: pyridine nucleotide dehydrogenases which oxidize α-hydroxycarboxylic acids at the α-position, all have A stereospecificity for the pyridine nucleotide, regardless of their stereospecificity for the substrate. Biellman and Rosenheimer [88] have assembled the data. One can add liver malic enzyme [90] to their list.

A word now about the stereospecificity for the substrate other than the pyridine nucleotide. Let us take lactic acid as an example. There are L-lactic dehydrogenases and D-lactic dehydrogenases, sometimes in the same cell. Why? The living cell regards L-lactate and D-lactate as quite different reagents, regardless of what the chemistry textbooks say. The L-lactic dehydrogenases have a function which is quite different from that of the D-lactic dehydrogenases. The enzymes are probably located in different places in the cell. Alizade and Simon [94] have demonstrated such compartmentation for a bacterium, with the use of labelled precursors in a quite ingeneous way. In the case of green leaves, it has been shown that D-lactic dehydrogenase is located in the organelles called peroxisomes [95]. The D-lactic dehydrogenase functions in the interconversion of D-glycerate and hydroxypyruvate, and of glyoxylate and glycolate, — not in the interconversion of pyruvate and lactate [95]. The autotrophically grown green alga Chlorella contains the same enzyme, and converts carbohydrate to D-lactate anaerobically in the dark [96]. But Chlorella cells cannot grow anaerobically in the dark on a mineral salts medium. The formation of D-lactate is probably not a normal or necessary component of their autotrophic metabolism. The D-lactate dehydrogenase of green leaves and algae should be called D-glycerate dehydrogenase. Its function has been discussed by Tolbert [95].

In contrast, L-lactic dehydrogenase functions in the anaerobic degradation of carbohydrate. D-triose phosphate is first oxidized by NAD to D-phosphoglycerate, which is converted to pyruvate, which reoxidizes the NADH formed in the prior redox reaction. To the best of my knowledge, it always seems to be L-lactic dehydrogenase which functions *physiologically* in this way. When alcohol dehydrogenase takes over the function of L-lactic dehydrogenase, the latter enzyme may drop out, as in green leaves.

In summary different stereospecificity of the enzymes suggests different physiological function and a different location in the cell.

The Structure of Lactic Dehydrogenase

Structural studies on L-lactic dehydrogenase are well advanced [97–99]. We probably have more information about this enzyme than we have about any other pyridine nucleotide dehydrogenase. Fig. *10* shows a diagram of the A-specific active site. The subunits of lactic dehydrogenase will hybridize, though they are from such different species as dog fish, chicken, beef and rabbit [100], so we assume all L-lactic dehydrogenases are very much alike. Structural studies on L-malic dehydrogenase are sufficiently far advanced to show that there are striking similarities in

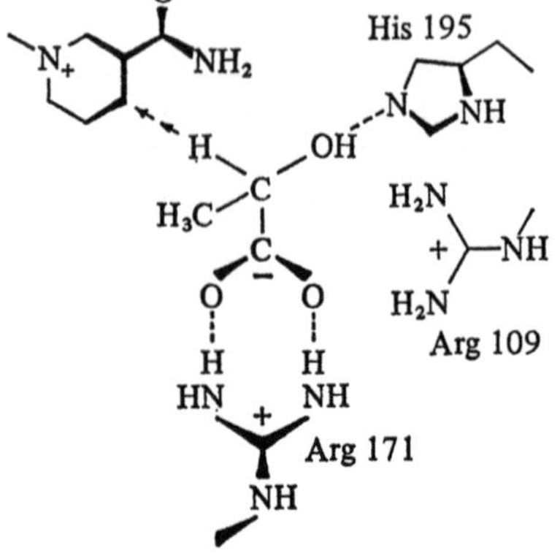

10

Diagrammatic representation of anticipated substrate binding in the active ternary intermediate of the lactic dehydrogenase reaction. The numbers refer to specific amino acid residues in the peptide chain of the protein.

[From Adams *et al.*, Proc. Natl. Acad. Sci. U.S. *70*, 1968 (1973), Fig. 5]

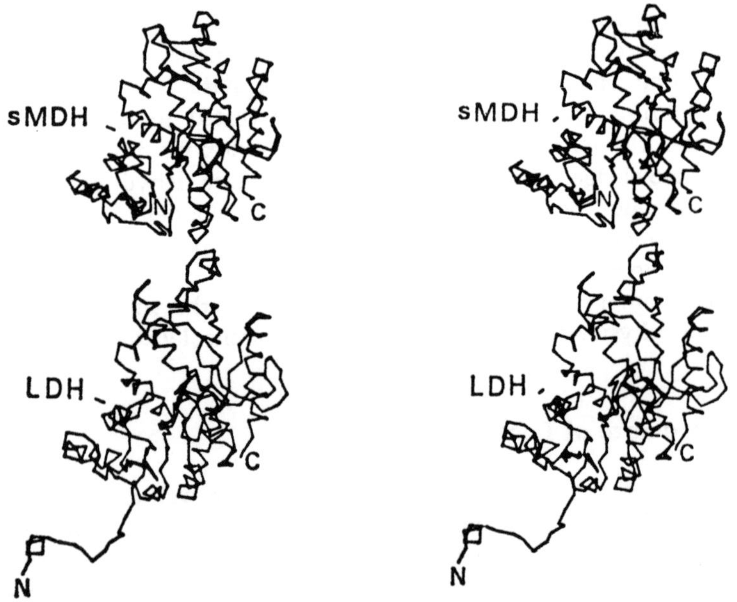

11

Stereoview of the α-carbon models of the subunits of s-malic dehydrogenase (sMDH) and lactic dehydrogenase (LDH).

[From Hill *et al.*, J. Mol. Biol. *72*, 577 (1972), Fig. 6] the tertiary structure of malic dehydrogenase and lactic dehydrogenase, as shown in Fig. 11 [101]. These are enzymes which both oxidize α-hydroxy carboxylic acids to α-oxo carboxylic acids, with transfer of hydrogen to the 4 A position of the nicotinamide ring of the pyridine nucleotide. Structural studies on triose-phosphate dehydrogenase, a B-specific enzyme, are also well under way [9,67]. It will be very interesting to see and compare an A-specific site and a B-specific site.

My generation was brought up strictly by its scientific parents. Free-wheeling speculation might be lots of fun, even necessary for scientific progress, but, like love, it wasn't supposed to be practiced in public. Standards of propriety have relaxed somewhat, and the new breed of "molecular" biologists speculate without shame before large and intensely interested audiences. There was a fear, before, that such a practice would result in science becoming somehow less scientific. It seems to me that the result has been exactly the opposite. I am not a molecular biologist, but more of an oldfashioned physiologist. A physiologist is someone who starts with a biological phenomenon that seems messy but also mysterious and cleans it up to a point where the chemists find it worthwhile to take over. A physiologist doesn't need to know much chemistry. Therefore, I should never have undertaken the responsibility for presenting biological stereochemistry to chemists, but there was this little thing I wanted to say.

In scientific prose the words chosen have to be given very precisely defined meanings. Therefore, no poetry please. The danger in poetry is that it will mean different things to different people. So people misunderstand each other. Scientists know all about the importance of clarity and precise definition and never cause semantic problems, or do they? Listen to the expression "science for its own sake". That was probably borrowed from the expression "art for its own sake". The impression it was originally intended to convey was one of devotion! The real artist is driven by an inner compulsion, and is willing to starve in his attic trying to produce art, whether you pay him for it or not (for a while at least). The real artist is also the artist who can't do anything well, if it isn't in accord with his inner drive. Of course you can compel him, but then he can't produce art. The scientist likes to think of himself in this way too. (Are there two cultures?) But scientists don't starve in attics. They live very well, thankyou. So new names are needed now for old concepts, or the original meaning of the old concepts will be completely forgotten.

Pure science is a term we should reserve for the aspect of science associated with the urge to know and understand, which is built into the structure of the mind of all of us. Pure science is selfless. It should not be personified! We can all practice it.

There is a growing conviction, for some maybe just a wild hope, that the structure of living things is so completely logical that it carries within itself a record of its past, so that if we only know enough about it, we will understand how it developed out of chaos, and be able to direct its future. At the same time there is the realization that the problem is so terribly complicated that its solution requires the combined imaginative power of many brains fortified by many computers. Ideas have often been regarded rather like pieces of property, as though they could be patented and sold for money. An idea in "pure" science is really something that has no value whatever, unless it is shared, and becomes the common property of all of us. That's what confers the purity on science. I don't know enough genetics to explain how the molecular biologists (with the X-ray crystallographers and protein chemists) are going about trying to solve what some want to call "the secret of life itself", but when I listen to them and try to imagine what their final solution is going to look like, it seems to me that there will be stereospecificity everywhere.

This is written in an unconventional manner, in tribute to van't Hoff, who was an unconventional man.

Acknowledgement. I am grateful to Dr. H. R. Levy for stimulating discussions which helped to bring me up to date in a field that I left some years ago, and to Dr. J. Jeffery for sending me manuscripts prior to publication.

References

1) Popjack, G.: Stereospecificity of enzymic reactions. In: The enzymes, 3rd edit. (ed. P. D. Boyer), Vol. II, p. 115—215. New York: Academic Press 1970.

2) Bentley, R.: Molecular asymmetry in biology, Vol. I. New York: Academic Press 1969.

3) Bentley, R.: Molecular asymmetry in biology, Vol. II. New York: Academic Press 1970.

4) Arigoni, D., Eliel, E. L.: Chirality due to the presence of hydrogen isotopes at noncyclic positions. In: Topics in stereochemistry, Vol. IV, p. 127. New York: John Wiley and Sons 1969.

5) Dunathan, H. C.: Stereochemical aspects of pyridoxal phosphate catalysis. Advan. Enzymol. *35*, 79—134 (1971).

6) Hess, G. P., Rupley, J. A.: Structure and function of proteins. Ann. Rev. Biochem. *40*, 1013—1043 (1971).

7) Hanson, K.R.: Enzyme symmetry and enzyme stereospecificity. Ann. Rev. Plant Physiol. *23*, 335—366 (1972).

8) Dickerson, R. E.: X-ray studies of protein mechanisms. Ann. Rev. Biochem. *41*, 815—842 (1972).

9) Cold Spring Harbor Symposia on Quantitative Biology. Vol. 36, Structure and Function of Proteins at the Three-Dimensional Level, 1972.

[10] Rose, I. A.: Enzymology of proton abstraction and transfer reactions. In: The enzymes, 3rd edit. (ed. P. D. Boyer), Vol. II, p. 281—333. New York: Academic Press 1970.

[11] Rose, I. A.: Enzyme reaction stereospecificity: A critical review. CRC Critical Rev. Biochem. *1972*, 33—57.

[12] Korman, E. F., McLick, J.: Stereochemical reaction mechanism formulations for enzyme-catalyzed pyrophosphate hydrolysis, ATP hydrolysis, and ATP synthesis. Bioorganic Chem. *2*, 179—190 (1973).

[13] Perutz, M. F.: X-ray analysis, structure and function of enzymes (The First Sir Hans Krebs Lecture). European J. Biochem. *8*, 455—466 (1969).

[14] Perutz, M. F.: Stereochemistry of cooperative effects in haemoglobin. Nature *228*, 726—739 (1970).

[15] Anfinsen, C. B.: The formation and stabilization of protein structure. Biochem. J. *128*, 737—749 (1972).

[16] Milstien, S., Cohen, S. A.: Rate acceleration by stereopopulation control: Models for enzyme action. Proc. Natl. Acad. Sci. U.S. *67*, 1143—1147 (1970).

[16a] Ogston, A. G.: Interpretation of experiments on metabolic processes, using isotopic tracer elements. Nature 162, 963 (1948).

[17] Palekar, A. G., Tate, S. S., Meister, A.: Rat liver aminomalonate decarboxylase. Identity with cytoplasmic serine hydroxymethylase and allothreonine aldolase. J. Biol. Chem. *248*, 1158—1167 (1973).

[17a] Westheimer, F. H., Fisher, H. F., Conn, E. E., Vennesland, B.: The enzymatic transfer of hydrogen from alcohol to DPN. J. Am. Chem. Soc. *73*, 2403 (1951).

[17b] Fisher, H. F., Conn, E. E., Vennesland, B., Westheimer, F. H.: The enzymatic transfer of hydrogen I. The reaction catalyzed by alcohol dehydrogenase. J. Biol. Chem. *202*, 687—697 (1953). Loewus, F. A., Ofner, P., Fisher, H. F., Westheimer, F. H., Vennesland, B.: The enzymatic transfer of hydrogen II. The reaction catalyzed by lactic dehydrogenase. J. Biol. Chem. *202*, 699—704 (1953). Loewus, F. A., Westheimer, F. H., Vennesland, B.: Enzymatic synthesis of the enantiomorphs of ethanol-1-d. J. Am. Chem. Soc. *75*, 5018—5023 (1953).

[18] Hirschmann, H.: The structural basis for the differentiation of identical groups in asymmetric reactions. In: Essays in biochemistry (ed. S. Graff), p. 156. New York: John Wiley & Sons 1956.

[19] Hirschmann, H., Hanson, K. R.: The differentiation of stereoheterotopic groups. European J. Biochem. *22*, 301—309 (1971).

[20] Cahn, R. S., Ingold, C. K., Prelog, V.: Specification of molecular chirality. Angew. Chem. *78*, 413—447 (1966). — Cahn, R. S., Ingold, C. K.: J. Chem. Soc. (London) *1951*, 612. — Cahn, R. S., Ingold, C. K., Prelog, V.: Experientia *12*, 81 (1956).

[21] Hanson, K. R.: Applications of the sequence rule. I. Naming the paired ligands g,g at a tetrahedral atom Xggij. II. Naming the two faces of a trigonal atom Yghi. J. Am. Chem. Soc. *88*, 2731—2742 (1966).

[22] IUPAC 1968 Tentative Rules, Section E, Fundamental Stereochemistry. Reprinted in European J. Biochem. *18*, 151—170 (1971).

[23] Biochemistry and Physiology of Visual Pigments, Bochum Symposium 1972 (ed. H. Langer). Berlin—Heidelberg—New York: Springer 1973.

[24] Horwitz, J., Heller, J.: Photoselection and linear dichroism of retinals. A method for identification and measurement of various geometrical isomers. J. Biol. Chem. *248*, 1051—1055 (1973).

[25] Lipmann, F.: The relation between the direction and mechanism of polymerization (ed. P. N. Campbell and G. D. Greville), Vol. IV. Essays Biochem. *1968*, 1—23.

26) Lipmann, F.: What do we know about protein synthesis? In: Gene expression and its regulation. New York: Plenum Publishing Corporation 1972.
27) Blundell, T., Dodson, G., Hodgkin, D., Mercola, D.: Insulin: the structure in the crystal and its reflection in chemistry and biology. Advan. Protein Chem. *26*, 279—402 (1972).
28) Beytia, E., Qureshi, A. A., Porter, J. W.: Squalene synthetase III: Mechanism of the reaction. J. Biol. Chem. *248*, 1856—1867 (1973).
29) Christopher, J. P., Pistorius, E. K., Regnier, F. E., Axelrod, B.: Factors influencing the positional specificity of soy bean lipoxygenase. Biochim. Biophys. Acta *289*, 82—87 (1972).
30) Hashimoto, H., Günther, H., Simon, H.: The stereochemistry of vinylacetyl-CoA-isomerase of *Clostridium kluyveri*. FEBS Letters *33*, 81—83 (1973).
31) Milborrow, B. V.: Stereochemical aspects of the formation of double bonds in abscisic acid. Biochem. J. *128*, 1135—1146 (1972).
32) Tsai, Su-Chen, Steinberg, D., Avigan, J., Fales, H. M.: Studies on the stereospecificity of mitochondrial oxidation of phytanic acid and of α-hydroxyphytanic acid. J. Biol. Chem. *248*, 1091—1097 (1973).
33) Richards, J. B., Hemming, F. W.: Dolichols, ubiquinones, geranylgeraniol and farnesol as the major metabolites of mevalonate in *Phytophthora cactorum*. Biochem. J. *128*, 1345—1352 (1972).
34) Egmond, M. R., Vliegenhart, J. F. G., Boldingh, J.: Stereospecificity of the hydrogen abstraction at carbon atom n-8 in the oxygenation of linoleic acid by lipoxygenases from corn germs and soya beans. Biochem. Biophys. Res. Commun. *48*, 1055—1060 (1972).
35) Reed, L. J., Cox, D. J.: Multienzyme complexes. In: The enzymes, 3rd edit. (ed. P. D. Boyer), Vol. I, p. 213—240. New York—London: Academic Press 1970.
36) Lipmann, F., Gevers, W., Kleinkauf, H., Roskoski, R., Jr.: Polypeptide synthesis on protein templates: The enzymatic synthesis of Gramicidin S and tyrocidine (ed. A. Meister). Advan. Enzymol. *35*, 1—34 (1971).
37) Seliger, H. H., McElroy, W. D., White, E. H., Field, G. F.: Stereospecificity and firefly bioluminescence, a comparison of natural and synthetic luciferins. Proc. Natl. Acad. Sci. U.S. *47*, 1129—1134 (1961).
38) White, E. H., McCapra, F., Field, G. F., McElroy, W. D.: The structure and synthesis of firefly luciferin. J. Am. Chem. Soc. *83*, 2402—2403 (1961).
38a) Cormier, M. J., Wampler, J. E., Hori, K.: Bioluminescence: Chemical aspects (ed. Herz, W., Grisebach, H. and Kirby, G. W.). Progr. Chem. Org. Nat. Prod. *30*, 1—60 (1973).
39) Eggerer, H., Buckel, W., Lenz, H., Wunderwald, P., Gottschalk, G., Cornforth, J. W., Donninger, C., Mallaby, R., Redmond, J. W.: Stereochemistry of enzymic citrate synthesis and cleavage. Nature *226*, 517—521 (1970).
40) O'Brien, R. W., Stern, J. R.: Reversal of the stereospecificity of the citrate synthase of *Clostridium kluyveri* by p-chloromercuribenzoate. Biochem. Biophys. Res. Commun. *34*, 271—276 (1969).
41) Cardinale, G. J., Abeles, R. H.: Purification and mechanism of action of proline racemase. Biochemistry *7*, 3970—3978 (1968).
42) Kosicki, G. W., Westheimer, F. H.: Oxaloacetate decarboxylase from cod. Mechanism of action and stereoselective reduction of pyruvate by borohydride. Biochemistry *7*, 4303—4309 (1968).
43) Phillips, T. M., Kosicki, G. W., Schmidt, D. E., Jr.: Stereoselective reduction of pyruvate by sodium borohydride catalyzed by pyruvate kinase. Biochim. Biophys. Acta *293*, 125—133 (1973).

44) Maxwell, E. S.: The enzymic interconversion of uridine diphosphogalactose and uridine diphosphoglucose. J. Biol. Chem. *229*, 139—151 (1957).
45) Seyama, Y., Kalckar, H. M.: Specific tritium labeling of uridine diphosphogalactose 4-epimerase by D-1-^{3}H Galactose. Biochemistry *11*, 36—39 (1972).
46) Seyama, Y., Kalckar, H. M.: Interaction between uridine diphosphate galactose and uridine diphosphate galactose 4-epimerase from *Escherichia coli*. Biochemistry *11*, 40—44 (1972).
47) Wee, T. G., Frey, P. A.: Studies on the mechanism of action of uridine diphosphate galactose 4-epimerase. Substate dependent reduction by sodium borohydride. J. Biol. Chem. *248*, 33—40 (1973).
48) Maitra, U. S., Ankel, H.: The intermediate in the uridine diphosphate galactose 4-epimerase reaction. J. Biol. Chem. *248*, 1477—1479 (1973).
49) Ketley, J. N., Schellenberg, K. A.: Substrate stereochemical requirements in the reductive inactivation of uridine diphosphate galactose 4-epimerase by sugar and 5'-uridine monophosphate. Biochemistry *12*, 315—320 (1973).
50) Eigen, M.: Selforganization of matter and the evolution of biological macromolecules. Naturwissenschaften *58*, 465—523 (1971).
51) Black, S.: A theory on the origin of life (ed. A. Meister). Advan. Enzymol. *38*, 193—234 (1973).
52) Woese, C. R.: The genetic code. New York–Evanston–London: Harper and Row 1967.
53) Crick, F. H. C.: The origin of the genetic code. J. Mol. Biol. *38*, 367—379 (1968).
54) Orgel, L. E.: Evolution of the genetic apparatus. J. Mol. Biol. *38*, 381—393 (1968).
55) Decker, P.: Evolution in open systems: Bistability and the origin of molecular asymmetry, Nature New Biology *241*, 72—74 (1973).
56) Segal, H. L.: On the origin of stereospecificity in biological systems. FEBS Letters *20*, 255—256 (1972).
57) Lacey, J. C. Jr., Pruitt, K. M.: Origin of the genetic code. Nature *223*, 799—804 (1969).
58) Blow, D. M., Birktoft, J. J., Hartley, B. S.: Role of a buried acid group in the mechanism of action of chymotrypsin. Nature *221*, 337—340 (1969).
59) Hartley, B. S.: The Evolution of Enzymes. Plenary Lecture, Ninth International Congress of Biochemistry, Stockholm, Abstracts, 7 (1973).
60) Margoliash, E., Fitch, W. M., Dickerson, R. E.: Molecular expression of evolutionary phenomena in the primary and tertiary structures of cytochrome c. Brookhaven Symp. Biol. *21*, 259 (1968).
61) Dickerson, R. E.: The structure of cytochrome *c* and the rates of molecular evolution. J. Mol. Evolution *1*, 26—45 (1971).
62) Dickerson, R. E., Takano, T., Eisenberg, D., Kallai, O. B., Samson, L., Cooper, A., Margoliash, E.: Ferricytochrome c. I. General features of the horse and bonito proteins at 2.8 A resolution. J. Biol. Chem. *246*, 1511—1533 (1971).
63) Smith, L., Davies, H. C., Reichlin, M., Margoliash, E.: Separate oxidase and reductase reaction sites on cytochrome c demonstrated with purified site-specific antibodies. J. Biol. Chem. *248*, 237—243 (1973).
64) Levy, H. R., Talalay, P., Vennesland, B.: The steric course of enzymatic reactions at meso carbon atoms: application of hydrogen isotopes. In: progress in stereochemistry (ed. de la Mare and Klyne), Vol. 3, p. 299—349. (London: Butterworths 1962.
65) Cornforth, J. W., Ryback, G., Popjack, G., Donninger, C., Schroepfer, G., Jr.: Stereochemistry of enzymic hydrogen transfer to pyridine nucleotides Biochem. Biophys. Res. Commun. *9*, 371 (1962).

66) Voet, J. G., Hindenlang, D. M., Blanck, T. J. J., Ulevitch, R. J., Kallen, R. G., Dunathan, H. C.: The stereochemistry of pyridoxal phosphate enzymes. The absolute stereochemistry of cofactor C'4 protonation in the transamination of holoserine hydroxymethylase by D-alanine. J. Biol. Chem. *248*, 841—842 (1973).
67) Pyridine nucleotide-dependent dehydrogenases (ed. H. Sund). Berlin—Heidelberg—New York: Springer 1970.
68) Harris, J. I.: The primary structure and activity of glyceraldehyde. 3-phosphate dehydrogenase, in Ref. 67, 57—70.
69) Adams, M. J., McPherson, A., Jr., Rossmann, M. G., Schevitz, R. W., Smiley, I. E., Wonacott, A. J.: Structure and mechanism of lactic dehydrogenase, in Ref. 67, 157—174.
70) Hung, H. C., Hoberman, H. D.: Influence of steric specificity on the rates of hydrogen exchange between substrates of NAD-coupled dehydrogenases. Biochem. Biophys. Res. Commun. *46*, 399—405 (1972).
71) Dugan, R. E., Porter, J. W.: Stereospecificity of the transfer of hydrogen from reduced nicotinamide adenine dinucleotide phosphate, in each of the two reductive steps catalyzed by β-hydroxy-β-methylglutaryl coenzyme A reductase. J. Biol. Chem. *246*, 5361—5364 (1971).
72) Beedle, A. S., Munday, K. A., Wilton, D. C.: The stereochemistry of hydrogen transfer from NADPH catalyzed by 3-hydroxy-3-methylglutaryl-coenzyme A reductase from rat liver. European J. Biochem. *28*, 151—155 (1972).
73) Blattmann, P., Rétey, J.: Zur Wirkungsweise und Stereospezifität der Hydroxymethylglutaryl CoA-Reduktase. Hoppe-Seylers Z. Physiol. Chem. *352*, 369—376 (1971).
74) Beedle, A. S., Munday, K. A., Wilton, D. C.: The stereochemistry of the reduction of mevaldic acid-coenzyme A hemithioacetal by rat liver 3-hydroxy-3-methylglutaryl coenzyme A-reductase. FEBS Letters *28*, 13—15 (1972).
75) Rétey, J., von Stetten, E., Coy, U., Lynen, F.: A probable intermediate in the enzymic reduction of 3-hydroxy-3-methylglutaryl coenzyme A. European J. Biochem. *15*, 72—76 (1970).
76) Seiler, M. P., Acklin, W., Arigoni, D.: Cited in Ref. 73.
77) Gang, H., Cederbaum, A. I., Rubin, E.: Stereospecificity of ethanol oxidation. Biochem. Biophys. Res. Commun. *54*, 264—269 (1973).
78) Gibb, W., Jeffery, J.: The steric course with respect to the reduced nicotinamide-adenine dinucleotide of the reduction of 3-oxo steroids catalyzed by cortisone reductase. European J. Biochem. *34*, 395—400 (1973).
79) Betz, G., Warren, J. C.: Reaction mechanism and stereospecificity of 20 β-hydroxysteroid dehydrogenase. Arch. Biochem. Biophys. *128*, 745—752 (1968).
80) Gibb, W., Jeffery, J.: Relationships between the 3 α and 20 β-hydroxysteroid NAD-oxidoreductase activity of a crystalline-enzyme preparation. European J. Biochem. *23*, 336—342 (1971).
81) Gibb, W., Jeffery, J.: Reduction of the non-steroid adamantanone by crystalline preparations of cortisone reductase. Biochem. J. *126*, 443 (1972).
82) Gibb, W., Jeffery, J.: Steric, chiral and conformational aspects of the 3-hydroxy- and 20-hydroxysteroid dehydrogenase activities of cortisone reductase preparations. Biochim. Biophys. Acta *268*, 13—20 (1972).
83) Gibb, W., Jeffery, J.: 5α-dihydrotestosterone sulfate and cortisone reductase. Biochim. Biophys. Acta *280*, 646—651 (1972).
84) George, J. M., Orr, J. C., Renwick, A. G. C., Carter, P., Engel, L. L.: The stereochemistry of hydrogen transfer to $NADP^+$ by enzymes acting upon stereoisometric substrates. Bioorganic Chem. *2*, 140—144 (1973).

85) Middleditch, L. E., Chung, A. E.: Pyridine nucleotide transhydrogenase from *Azotobacter vinelandii* cells: Stereospecificity of hydrogen transfer. Arch. Biochem. Biophys *146*, 449—453 (1971).

86) Schutzbach, J. S., Feingold, D. S.: Biosynthesis of uridine diphosphate D-xylose. IV. Mechanism of action of uridine diphosphoglucuronate carboxy-lyase. J. Biol. Chem. *245*, 2476—2482 (1970).

87) Rose, I. A.: Stereochemistry of pyruvate kinase, pyruvate carboxylase, and malate enzyme reactions. J. Biol. Chem. *245*, 6052—6056 (1970).

88) Biellmann, J.-F., Rosenheimer, N.: Dogfish lactate dehydrogenase: The stereochemistry of hydrogen transfer. FEBS Letters *34*, 143—144 (1973).

89) Davies, D. D., Teixeira, A., Kenworthy, P.: The stereospecificity of nicotinamide-adenine dinucleotide-dependent oxido-reductases from plants. Biochem. J. *127*, 335—343 (1972).

90) Krakow, G., Ludowieg, J., Mather, J. H., Normore, W. M., Tosi, L., Udaka, S., Vennesland, B.: Some stereospecificity studies with tritiated pyridine nucleotides. Biochemistry *2*, 1009—1014 (1963).

91) Biellmann, J.-F., Branlant, G., Olomucki, A.: Stereochemistry of the hydrogen transfer to the coenzyme by octopine dehydrogenase. FEBS Letters *32*, 254—256 (1973).

92) Fisher, H. F., Adija, D. L., Cross, D. G.: Dehydrogenase-reduced coenzyme difference spectra, their resolution and relationship to the stereospecificity of hydrogen transfer. Biochemistry *8*, 4424—4430 (1969).

93) Agranoff, B. W., Hajra, A. K.: The acyl dihydroxyacetone phosphate pathway for glycerolipid biosynthesis in mouse liver and Ehrlich ascites tumor cells. Proc. Natl. Acad. Sci. U.S. *68*, 411—415 (1971).

94) Alizade, M. A., Simon, H.: Studies on mechanism and compartmentation of the L- and D-lactate formation from L-malate and D-glucose by *Leuconostoc mesenteroides*. Hoppe-Seylers Z. Physiol. Chem. *354*, 163—168 (1973).

95) Tolbert, N. E.: Microbodies-peroxisomes and glyoxysomes. Ann. Rev. Plant Physiol. *22*, 45—74 (1971).

96) Warburg, O., Gewitz, H.-S., Völker, W.: D, Milchsäure in Chlorella. Z. Naturforsch. *12b*, 722—724 (1957).

97) Everse, J., Kaplan, N. O.: Lactate dehydrogenases: Structure and function (ed. A. Meister). Advan. Enzymol. *37*, 61—131 (1973).

98) Adams, M. J., Ford, G. C., Koekoek, R., Lentz, P.J., Jun. McPherson, A., Jun., Rossmann, M. G., Smiley, I. E., Schevitz, R. W., Wonacott, A. J.: Structure of lactate dehydrogenase at 2.8 A resolution. Nature *227*, 1098—1103 (1970).

99) Adams, M. J., Buehner, M., Chandrasekhar, K., Ford, G. C., Hackert, M. L., Liljas, A., Rossmann, M. G., Smiley, I. A., Allison, W. S., Everse, J., Kaplan, N. O., Taylor, S. S.: Structure-function relationships in lactate dehydrogenase. Proc. Natl. Acad. Sci. U.S. *70*, 1968—1972 (1973).

100) Chilson, O. P., Costello, L. A., Kaplan, N. O.: Studies on the mechanism of hybridization of lactic dehydrogenases *in vitro*. Biochemistry *4*, 271—281 (1965).

101) Hill, E., Tsernoglou, D., Webb, L., Banaszak, L.: Polypeptide conformation of cytoplasmic malate dehydrogenase from an electron density map at 3.0 A resolution. J. Mol. Biol. *72*, 577—591 (1972).

Received October 5, 1973

[2.2]Paracyclophanes, Structure and Dynamics*

Prof. Dr. Fritz Vögtle and Dr. Peter Neumann

Institut für organische Chemie der Universität, Würzburg

Contents

* Chemistry of Phanes, Comm. 5. — Comm. 4: Vögtle, F., Neumann, P.: Synthesis *1973*, 85.

Summary

Non-planar benzene rings, anomalous bond lengths and angles, and electronic interactions between parallel benzene nuclei are common features of the [2.2]paracyclophane system. This progress report reviews the unusual physical and, in particular, spectroscopic properties of such a strained molecular structure. Chirally substituted [2.2]paracyclophanes offer parallels to the stereochemistry of the metallocenes. Characteristics of the chemical behaviour of [2.2]paracyclophanes and analogous compounds comprise: transannular directing effects in electrophilic substitution, neighboring-group effects of the [2.2]paracyclophanyl moiety, cis-addition at the aliphatic bridges, dynamic intramolecular processes such as isomerization and racemization, and photochemical reactions.

1. Introduction

The properties of the cyclophanes are best illustrated by the paracyclophanes. In contrast to the metacyclophanes [1] and metaparacyclophanes [2], where aromatic nuclei come into close proximity, there are in paracyclophane molecules two aromatic nuclei pressed one on top

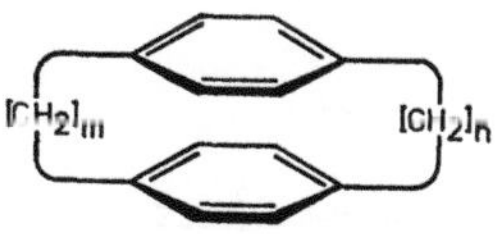

1

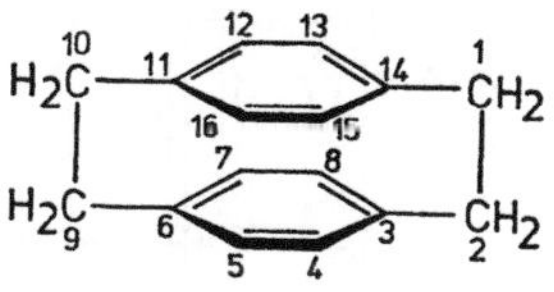

2

of the other in parallel planes. In the homologous series of [m.n]paracyclophanes (*1*) the initial member, *i.e.* the member containing the shortest possible bridge, is [2.2]paracyclophane (*2*). The boat-shaped form of the benzene rings and the inter-electronic effects between the spatially penetrating π electron clouds in such molecules are interesting objects for both theoretical and experimental study. Since X-ray structural analyses have provided a firm conceptual basis for spatial models and promising theoretical (MO) approaches are available, to seek an exact interpretation of the transannular interactions and their mechanisms is a highly topical research problem, as this report shows.

2. Molecular Geometry and Physical Properties

2.1. [2.2]Paracyclophane

The synthesis of [2.2]paracyclophane (*2*) and its identification by X-ray structural analysis were first reported in a short communication by Brown and Farthing [3] in 1949. A more detailed report on its molecular structure followed in 1953[4]. Further investigations by Lonsdale *et al.*[5] at two different temperatures (93 and 291 °K) provided additional information about the thermal expansion and molecular vibrations in the crystal. A recent X-ray structural analysis [6] confirms and supplements Lonsdale's observations.

According to Lonsdale *et al.*, the four aromatic bridgehead atoms C_3, C_6, C_{11} and C_{14} are bent out of the plane of the remaining benzene carbon atoms by about 0.168 Å (14°) at 291 °K (see Fig. 1); the aromatic nuclei are deformed into a boat conformation. The distance between the aromatic bridgehead atoms C_3, C_{14} and C_6, C_{11} has shrunk to 2.751 Å. The intramolecular distance between the plane formed by atoms C_4, C_5, C_7, C_8 and that formed by atoms C_{12}, C_{13}, C_{15}, C_{16} is only 3.087 Å; the van der Waals distance between two parallel benzene nuclei is usually at least 3.4 Å. There must therefore be considerable transannular π-overlapping in compound *2*. To compensate for this the CH_2–CH_2 bond length is unusually large: 1.630 Å at 291 °K; at 93 °K it is only 1.558 Å.

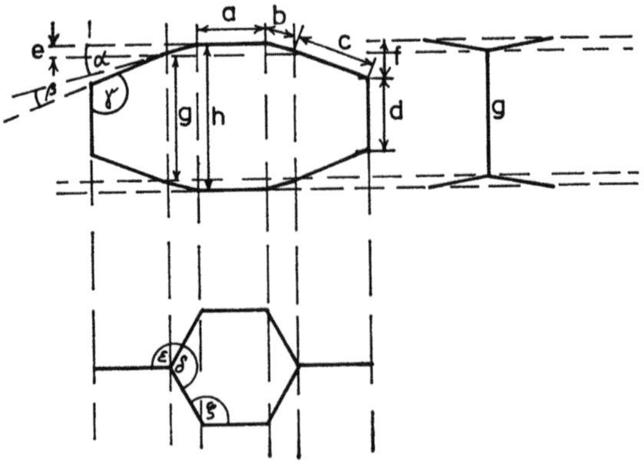

Fig. 1. Geometry of the [2.2]paracyclophane molecule in the crystal. (For numerical values see Table 1)

The strain energy of the molecule, estimated by Boyd [7–9] to be 31 kcal/mol, is obviously not confined to part of the molecule but distributed over the whole molecular skeleton [10]. Attribution of this strain energy to the various structural elements shows that the greatest contribution comes from the distortion of the benzene nucleus away from a planar arrangement.

Starting from empirical valence force potentials, Boyd [8] has calculated the lowest-energy conformation of [2.2]paracyclophane. The geometrical data are in satisfactory agreement with experimentally determined values (see Table 1).

It is noteworthy that intramolecular steric interactions still distort the molecule, albeit less so, in the case of [3.3]paracyclophane (3) [11]. Here the strain energy is only 7 kcal/mol, mostly contributed by the

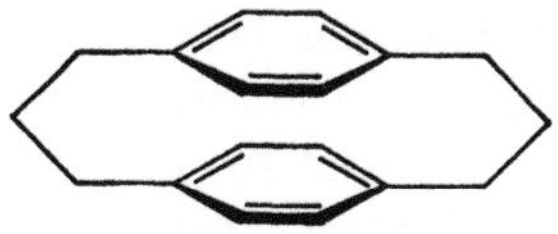

3

distortion of the aromatic ring away from a planar conformation. The angle α through which the bridgehead atoms are displaced from the plane of the other four aromatic carbon atoms is 6.4° (see Fig. 2), and

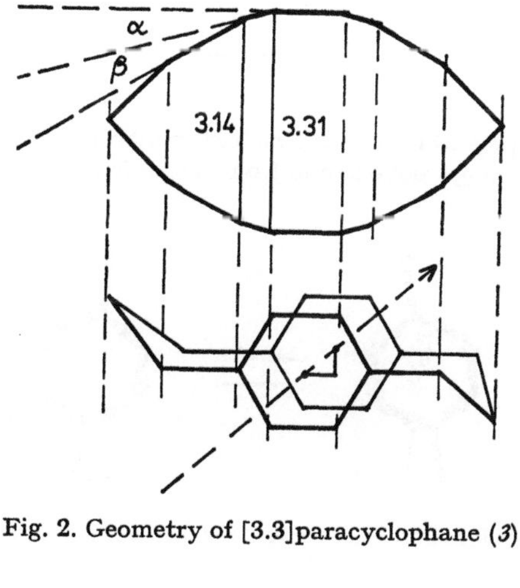

Fig. 2. Geometry of [3.3]paracyclophane (*3*)

angle β only 2.5 or 4.6°. The values for β are different because the benzene rings are not coplanar but about 0.5 Å apart, though still parallel. Strain is also reduced in [3.3]paracyclophane because of its longer bridges. The transannular distance of the aromatic bridgehead atoms has increased to 3.14 Å and the distance between the central carbon atoms (3.29 and

3.31 Å) almost equals the van der Waals distance. As in [2.2]paracyclophane, the hydrogen atoms of the aromatic moieties are directed towards the interior of the molecule because of the increased π-electron density on the outside of the molecule caused by the intraannular interaction of the π-electron clouds.

The outcome is a repulsion of the C—H σ-bond electrons. The methylene carbon atoms in [3.3]paracyclophane assume a chairlike conformation; in solution the chair and boat forms were found to be in equilibrium [12]. The slight distortion of the molecule is reflected in the expansion of the bond angles and in dihedral angles at the bridges. The lengths of the aliphatic C—C bonds are normal (1.507 and 1.517 Å).

Lonsdale *et al.*[5] concluded from data obtained at 93 and 291 °K that the [2.2]paracyclophane molecule vibrates in a concertina-like fashion, *i.e.* the benzene rings move to and fro perpendicular to their axes (see Fig. 3). Furthermore, simultaneous rotational movement (twisting) of the benzene rings with respect to each other takes place (see Fig. 4).

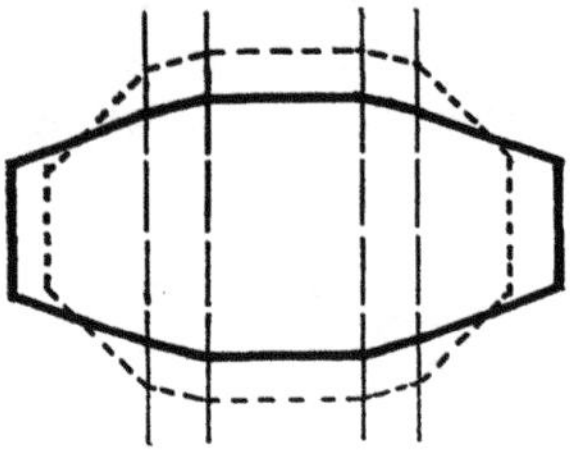

Fig. 3. Schematic representation of the molecular vibration of [2.2]paracyclophane, in which the benzene rings move to and fro in a concertina-like fashion with respect to each other

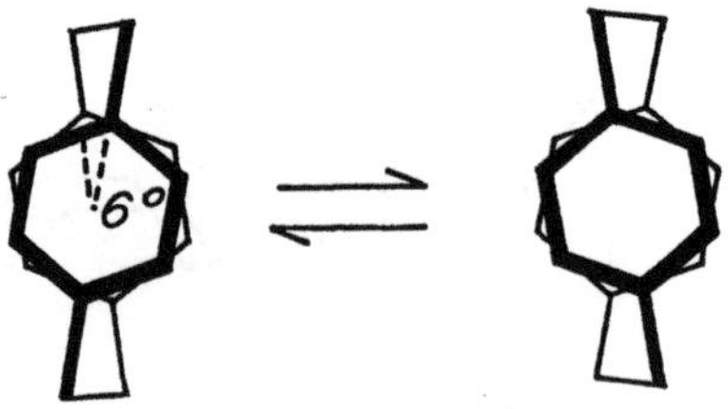

Fig. 4. Twisting vibration of the two benzene nuclei about the normal axis

These vibrations also occur, albeit with diminished amplitude, at lower temperatures (93 °K). According to Trueblood *et al.*[6] the angle of torsion is 6°. The structural data obtained by these authors are listed in Table 1 together with those reported by Lonsdale and Brown; the values calculated by Boyd are included for comparison.

Table 1. Bond lengths and bond angles in [2.2]paracyclophane

Bond width [in Å]	Brown [3]	Lonsdale [5] 291 °K	93 °K	Trueblood [6] 93 °K	Boyd [8]
a	1.400	1.421	1.415	1.387	1.390
b	1.390	1.380	1.388	1.384	1.390
c	1.540	1.547	1.534	1.512	1.548
d	1.548	1.630	1.558	1.562	1.570
e	0.133	0.168	0.172	0.1575 [1])	0.1385 [1])
f	0.771 [1])	0.728	0.768	0.7655 [1])	0.742 [1])
g	2.83	2.751	2.749	2.778	2.777
h	3.09	3.087	3.093	3.093	3.054

Bond angle

α	11° [2])	14.0°	14.0°	12.6°	11.8°
β	14° [2])	7° [2])	—	11.2°	11.5°
γ	114° 37′	111.2°	112.9°	113.7°	112.7°
δ	118° 36′	119.7°	118.4°	—	—
ε	119° 55′	119.9°	120.4°	—	—
ζ	120° 14′	119.1°	119.8°	—	—

[1]) Calculated from quoted data.
[2]) Calculated [8]).

Table 2. Comparison of the UV absorption of [2.2]paracyclophane with that of the model open-chain compound bis(p-ethylphenyl)butane

[2.2]Paracyclophane λ (nm)	log ε	Bis-(p-ethylphenyl)butane λ (nm)	Shift $\Delta\lambda$ (nm)
—	—	214	—
225	4.38	219	6
244 [1])	3.52 [19])	223	21
—	—	259	—
286	2.41	265	21
—	—	267	—
302 [1])	2.19 [19])	273	29

[1]) Shoulder.

The ^{1}H—NMR spectrum of [2.2]paracyclophane [13–18]) consists of two singlets of equal intensity for the methylene ($\tau = 6.96$) and aromatic protons; because of the shielding effect of the two benzene rings, the latter absorb at comparatively high field strength ($\tau = 3.70$).

The electronic spectra of the [2.2]paracyclophanes provide valuable information regarding the extent and mechanisms of transannular electronic interactions in strained π-electron systems. A rigid system like the [2.2]paracyclophane molecule is of great value as a model for checking theoretical data with a view to interpretating UV spectra of sterically hindered molecules ("overcrowded compounds").

The UV spectrum of [2.2]paracyclophane [19–22], compared with the spectra of model acyclic (see Table 2) and higher cyclic compounds, is characterized by the absence of fine structure, the occurrence of bathochromic shifts, and reduced extinction coefficients.

Two effects are generally held responsible for the bathochromic shift ($\Delta\lambda$) of *2* [21,22]:

1. Deviation from planarity of the benzene rings; this is manifested in the absorption at $\lambda = 286$ nm
2. Transannular electronic interactions ("inter-ring π overlap") would also appear to play a role, as evidenced by the absorption at $\lambda = 244$ nm.

Both effects are reflected in the shift of the 302-nm absorption band.

The fluorescence and absorption spectra of [2.2]paracyclophane have been investigated and interpreted by a number of authors [23–26]. Theoretical calculations of the energy levels with the aid inter alia of an exciton model are in agreement with spectroscopic findings. The methods of calculation are, of course, based on a π approximation and neglect the σ skeleton.

A deeper study of the electron spectrum and the ESR spectrum as well as of the photochemical behavior of *2* was made by Gleiter [27]. He showed that a σ-π separation in the [2.2]paracyclophane system is not indicated, since the 1,2 and 9,10 bonds in *2* are parallel to the p orbitals of the benzene rings and favorably situated for σ–π interaction.

El-Sayed [28] has reported on the phosphorescence spectrum of [2.2]paracyclophane. The emission differs both in wavelength (≈4700 Å) and in duration (3.3 s) from that of benzene (≈3400 Å, 6 s); hence a favorable "intersystem crossing" from the lowest singlet to the emitting triplet state was inferred. The emission spectrum also indicates that interactions take place between the two aromatic nuclei in the triplet state.

Investigations on the electron spin resonance of the radical anions [29–31] of [2.2]-and higher [n.n]paracyclophanes have shown that delocalization of the unpaired electrons over both aromatic nuclei is scarcely possible until the number of bridge members $n > 3$ [31]. In open-chain compounds of the type Ar-$[CH_2]_n$-Ar the corresponding condition is $n > 1$. This would suggest that the mechanism responsible for the transfer of electrons between the two benzene rings in bridged compounds differs from that in unbridged compounds.

[2.2]Paracyclophane also has a compact structure in solution, as is readily seen on comparing its partition coefficient with that of p-xylene for the system octanol/water [32]. Rigid superposition of the benzene rings leads to an intramolecular delocalization of the hydrophobic π-electron clouds and hence to an increased affinity for the aqueous phase. Accordingly, the logarithm of the partition coefficient is found to be smaller than the value observed for p-xylene, and not twice as large, as would be expected for completely hydrophobic surfaces.

2.2. Substituted [2.2]Paracyclophanes

2.2.1. Substitution at the Bridge

In the following section a few representatives from the series of [2.2]paracyclophanes bearing bridge substituents will be described. This type of compound has so far not been studied to the same extent as the [2.2]paracyclophanes having substituents in the nucleus (see Section 2.2.2.).

During the synthesis of [2.2]paracyclophanediene, Dewhirst and Cram [33] prepared the bis-geminal dibromides *4a* and *4b*, which were subsequently converted into the diketones *5a* and *5b*. The UV spectra of the bromides *4a* and *4b* show maxima at 236 nm. A comparison with the λ_{max} values of other [2.2]paracyclophanes brominated at the bridges suggests that the absorption at 236 nm is the 235 nm band of [2.2]paracyclophane, shifted bathochromically by the inductive effect of the

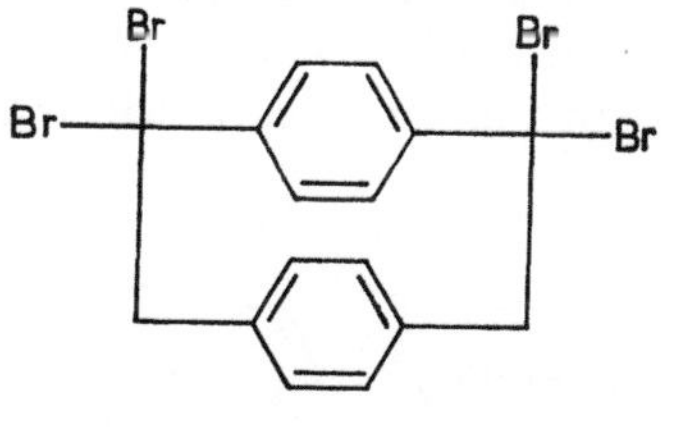

4a

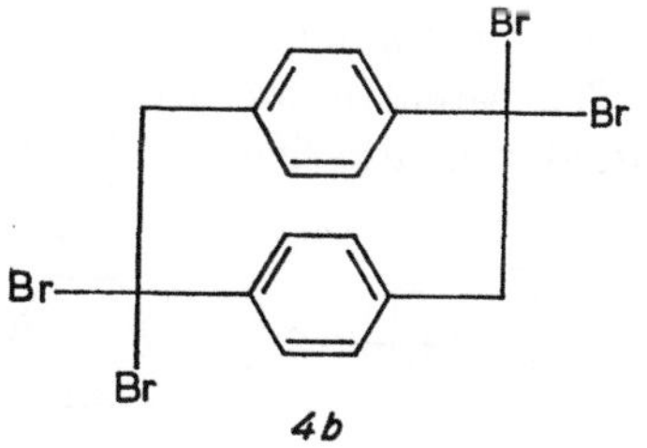

4b

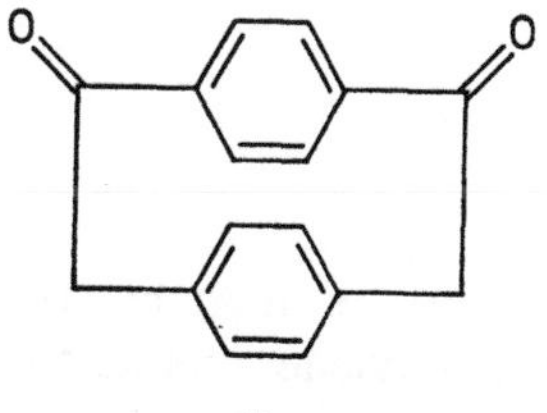

5a

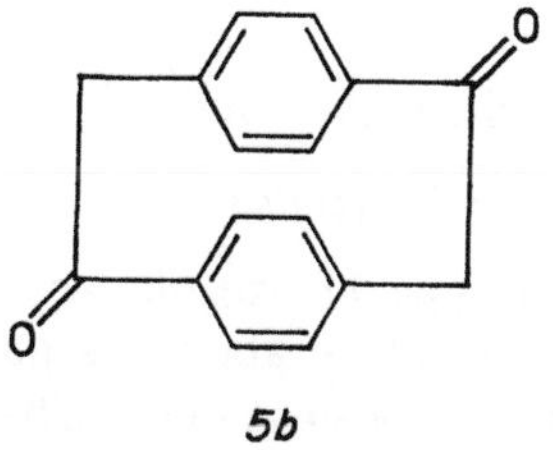

5b

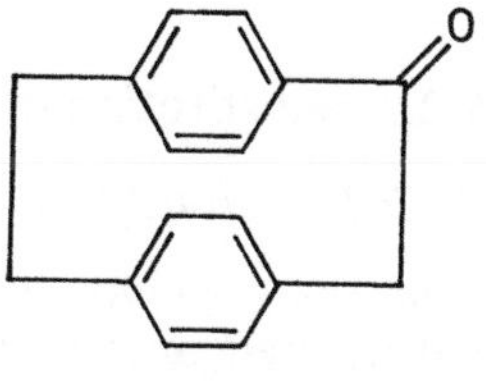

6

bromine atoms. The carbonyl groups in the diones *5a* and *5b*, whose maxima occur at 235 nm, probably have the same effect, though the position and intensity of the bands are consistent with the absorption of a relatively hindered benzoyl chromophore.

The infrared carbonyl frequencies of the monoketone (*6*), like those of *5a* and *5b*, appear at about 1700 cm^{-1} [17].

The spectroscopic properties (*e.g.* UV spectra) of the bridge-fluorinated 1,1,2,2,9,9,10,10-octafluoro[2.2]paracyclophane (*7*) resemble those of other [2.2]paracyclophanes [34].

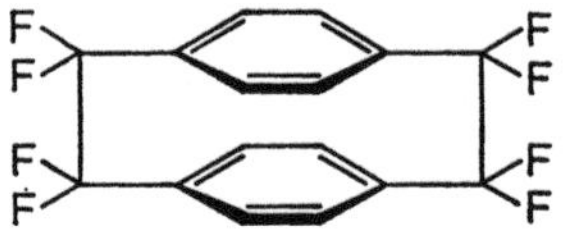

7

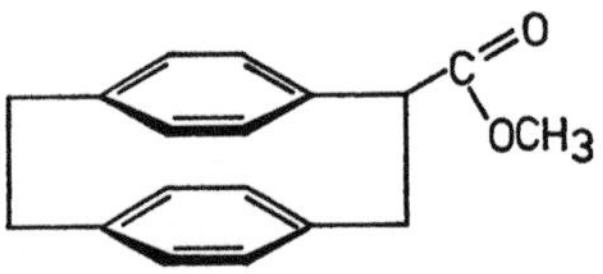

8

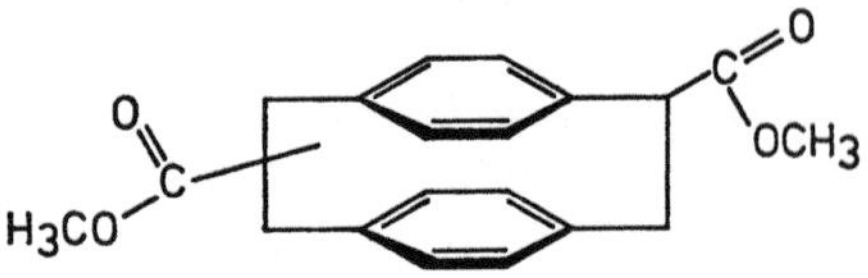

9

As expected, the ^1H–NMR spectrum shows only one absorption in the aromatic region (τ =2.7). This is shifted considerably compared with that of unsubstituted [2.2]paracyclophane (τ =3.7). In the IR spectrum an intense band occurs at 718 cm^{-1}, as is characteristic for other [2.2]paracyclophanes, *e.g.* methoxycarbonyl[2.2]paracyclophanes *8* and *9* [35], though it is absent in the spectrum of the linear poly-p-xylene (*10*). The parent hydrocarbon *2* shows this absorption, ascribed to the distorted aromatic rings, at 725 cm^{-1} [36].

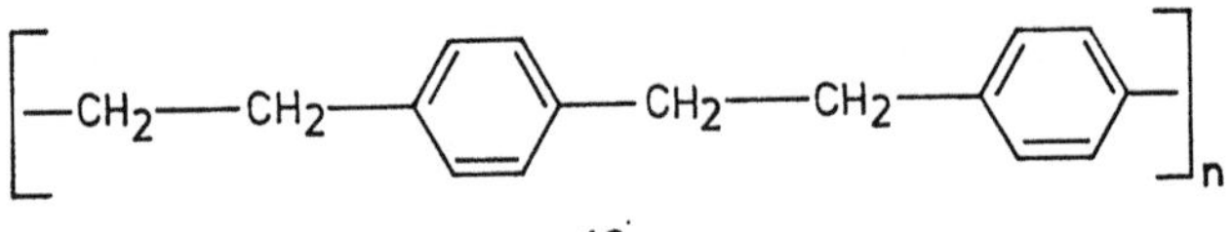

10

2.2.2. Substitution in the Nucleus

2.2.2.1. Chirality and Optical Activity

[m.n.]Paracyclophanes of type *11* are planar-chiral if, as assumed, the substituted nucleus cannot rotate about its paraphenylene axis at the temperatures at which it has been studied. The torsional stereoisomeric paracyclophanecarboxylic acids *11a* [37] and *11b* [38] have been resolved into their antipodes, whereas the cyclic compound *11c* with its greater number of bridge members cannot be resolved into the enantiomers

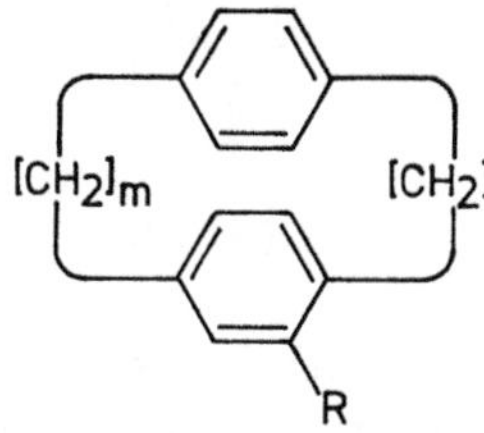

11a: R=COOH; *m*=*n*=2
11b: R=COOH; *m*=4; *n*=3
11c: R=COOH; *m*=*n*=4

at room temperature [39]. An analogous dependence of optical stability on the number of bridge members had earlier been established by Lüttringhaus for the ansa compounds ([m]paracyclophanes) of type *12* [40].

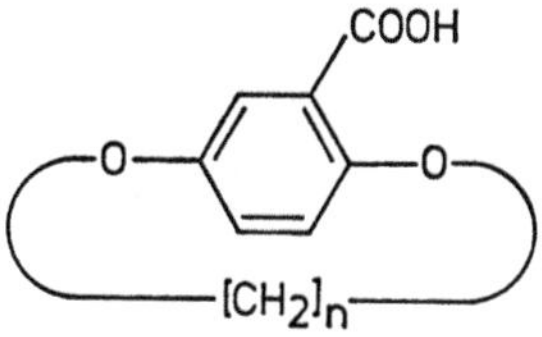

12

After they had successfully separated [2.2]paracyclophane-4-carboxylic acid *11a* [41] into its antipodes by fractional recrystallization of the (-)-α-phenylethylamine salt from ethanol, Cram and Allinger [37] turned their attention to the relative configurational arrangements within individual members of this class of compounds and the type of relationship existing between optical properties and absolute configuration.

This matter will now be dealt with in more detail since the theoretical and experimental methods and results of this aspect of phane chemistry are of fundamental importance in modern stereochemistry, as well as illustrating its scope.

Falk and Schlögl [42] as late as 1968 established the absolute configuration of compound *13a* as (+)−(S)-4-carboxy[2.2]paracyclophane. They did this by kinetic resolution of the racemic carboxylic anhydride with (−)-α-phenylethylamine: this kinetically controlled amidation afforded the dextrorotatory compound (*13*) in 3.8% optical yield [42]. The similar topology of the carboxyl-group environment in *13* and in α-substituted metallocene carboxylic acids (*14*), configurations con-

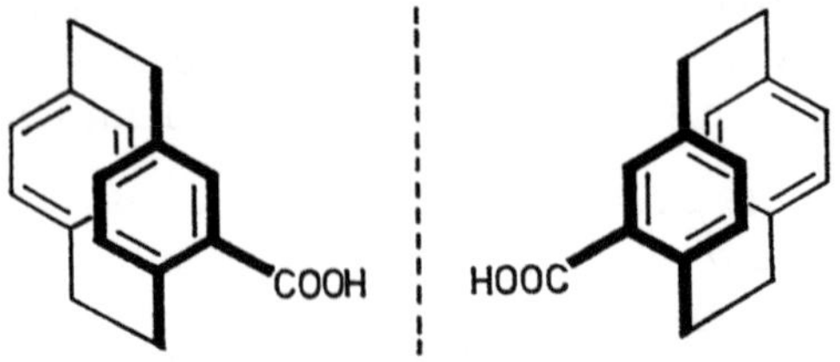

(+)-(S)-*13a* (-)-(R)-*13a*

firmed by X-ray structural analysis, permitted a direct comparison: (+)−*13* must have the same configuration (namely (+)−(S)) as (+)−*14*, which was also obtained by reaction of its anhydride with (−)-α-phenylethylamine.

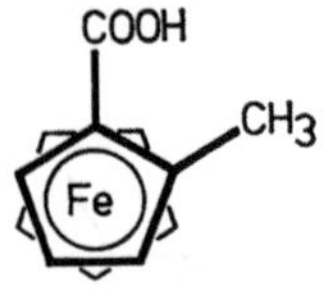

(+)−(S)−*14*

Another, independent method for establishing absolute configuration is to determine the configuration of the centers of chirality, as in *15* and *16*. The configuration of the asymmetric carbinol C atom is clearly linked with the planar-chiral part of the molecule; thus, if the position of the OH groups relative to the cyclophane residue (exo or endo) is known, the overall configuration can be deduced from the configuration of the center of chirality. This principle has already been successfully used in determining the configuration of the structurally analogous and optically active metallocenes [43].

The configuration of the carbinol C atom was determined as follows:

Reaction of exo-carbinol (−)−*16* with racemic α-phenylbutyric anhydride afforded the dextrorotatory acid; hence the chirality center has the absolute configuration (R). The behavior of the endo isomer *15* was also in accord with this finding: the resolved levorotatory acid indicates that the asymmetric C atom has the absolute configuration (S).

The planar chirality of (+)−*15* and (−)−*16* therefore has the absolute configuration (S), *i.e.* the results obtained by both methods agree on the absolute configuration of (−)−*13*.

The absolute configuration of more than 20 optically active [2.2]paracyclophane derivatives has been established in this way.

13

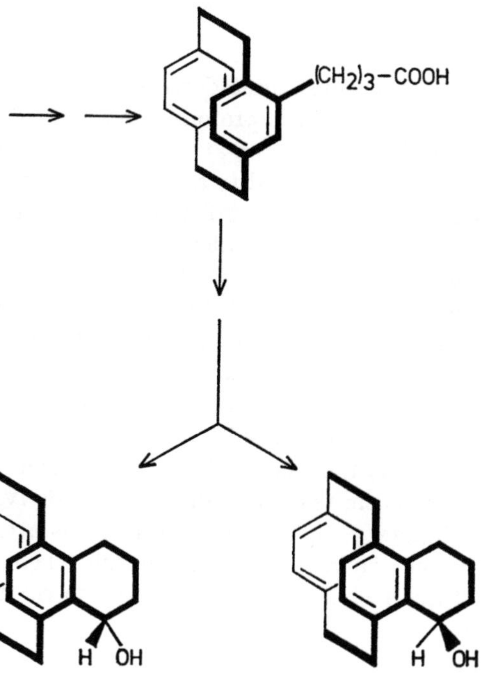

(+)-*15* (endo) (−)-*16* (exo)

The absolute configuration assigned to *13* on the basis of experimental findings has been theoretically confirmed by Weigang and Nugent[44]. These authors used a semi-empirical exciton theory for interpreting the optical activity of [2.2]paracyclophane derivatives; ring deformation and intra-annular overlap were, however, neglected in the calculations.

The circular dichroism of monosubstituted [2.2]paracyclophanes has been studied both theoretically and experimentally [45]. Theory suggests and experiment confirms that the sign of the Cotton effect of the first band ($^{b}L_1 \leftarrow A$) depends on the sign of the spectroscopic moment of the substituent in question. In the case of the following two bands, the signs of the Cotton effects are specific for the absolute configuration, regardless of substituent moment. Application of the octant rule to the n—π* transition of the aryl C=O chromophore for determining the absolute configuration in the [2.2]paracyclophane system has also been checked experimentally, which, however, requires separation of the n—π* effect from the „cyclophane background" (effect of the cyclophane skeleton itself). As other difficulties are also encountered, the use of circular dichroism in determination of cyclophane structure is problematical.

2.2.2.2. *Further Spectroscopic Findings*

Cram *et al.*[15,46] have investigated the ^1H–NMR spectra of [2.2]paracyclophane and nuclear-substituted derivatives.

A large shift of all aromatic protons was observed in the transition from electron-repelling to electron-attracting substituents. The proton ortho to the substituent is shifted to a higher field strength by electron-repelling substituents and to a lower field strength by electron-attracting substituents. The center of absorption of all the other protons in the aromatic moiety is shifted in the same direction as the ortho proton. The fact that the protons of the unsubstituted p-phenylene ring in monosubstituted [2.2]paracyclophanes are also shifted has been attributed to transannular electronic interactions [46].

The magnitude of the above-mentioned shifts of the ortho absorptions by various substituents (ortho shift) lies between the relatively large shifts of the hydrogen atoms cis to the substituents in vinyl compounds and the smaller ortho shifts of the usual benzene derivatives. The ortho shifts in the [2.2]paracyclophane system are thus attributed to an increased double-bond character in the deformed benzene rings, where canonical structures such as *17* could possibly contribute to stabilization of the molecule.

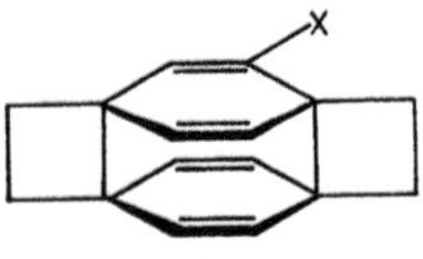

17

The IR spectrum of the "pseudo-geminal"[46] 4-acetyl-13-bromo[2.2]paracyclophane (*18*) shows a band for the carbonyl stretching vibration at 1663 cm^{-1}. This frequency lies outside the range of frequencies (1666—1668 cm^{-1}) found for the absorption of other isomers and has been attributed by Reich and Cram to transannular Br...C=O interactions.

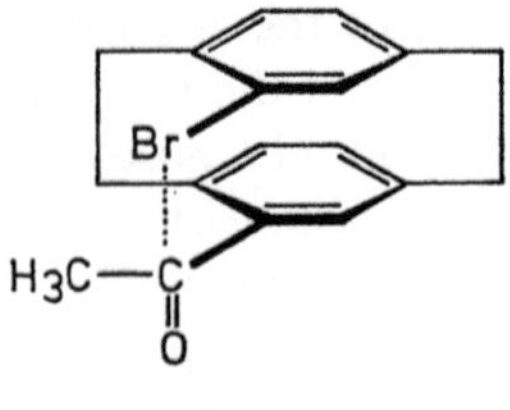

18

The mass spectra of the substituted [2.2]paracyclophanes at low ionizing voltages show distinct peaks for the chief radical fragments of the two p-xylylene halves of the molecule. Mass spectrometry thus provides a convenient tool for determining the nature and number of the substituents on each of the aromatic rings [46].

The ESR spectra of unusually stable nitroxide radicals such as *19* have been described by Forrester and Ramasseul [47].

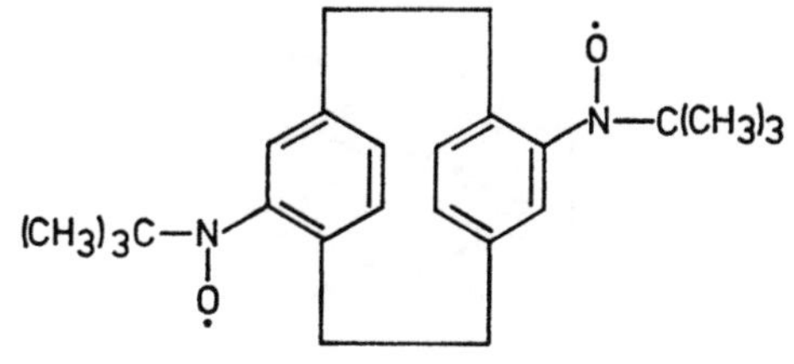

19

2.2.2.3. Tetra- and Octa-substituted [2.2]Paracyclophanes

4,7,12,15-Tetramethyl[2.2]paracyclophane (*20a*), in which the methyl groups are not in an "eclipsed" position, occurs as main product in the dimerization of 2,5-dimethyl-p-xylylene (*21*)[48]; the isomer (*20b*) was discovered much later [49].

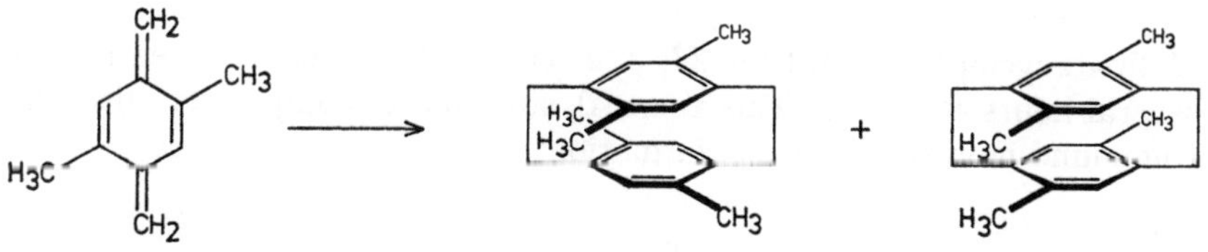

21 *20a* *20b*

20a is chiral and can be partially separated into the antipodes via the diastereomeric π complex with (—)—α-(2,4,5,7-tetranitro-9-fluorenylideneaminoxy)propionic acid [50]. The aromatic protons in *20b* are more strongly shielded than the corresponding hydrogen nuclei in *20a*. This has been attributed to a steric compression effect of the pseudo-geminal methyl groups [46].

Unlike other [2.2]paracyclophanes (see below), *20a* exhibits remarkable chemical stability. It is inert towards bromine in carbon tetrachloride, permanganate solution, and maleic anhydride, and cannot be reduced with Pd/C or nickel at room temperature under 1 atm hydrogen pressure.

In sharp contrast to *20a*, 4,5,7,8,12,13,15,16-octamethyl[2.2]paracyclophane *22* is extraordinarily unstable [51]. This substance polymerizes at room temperature, both in solution and in the solid state, even in an inert atmosphere. The reason is that the accumulation of pseudo-geminal methyl groups leads to steric overcrowding which cannot be circumvented because of the rigidity of the [2.2]paracyclophane system. Accordingly,

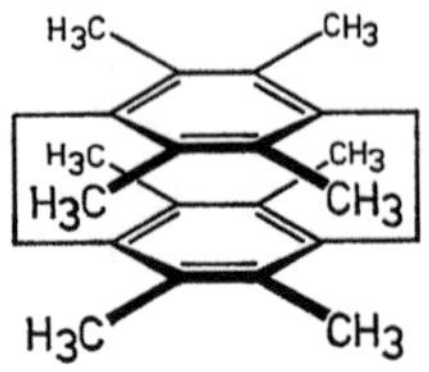

22

22 shows a smaller bathochromic shift (about 8 nm) and a significantly larger extinction coefficient above 300 nm than does *20a*. This is in complete agreement with the assumption that the longer-wavelength bands are associated with benzene ring deformation, while the shorter wavelengths are due to transannular electronic effects. Enhanced deformation of the benzene ring is, of course, to be expected in *22* because the steric interaction of the eight methyl groups is stronger than that of the transannular interactions.

Hopf [52] has recently synthesized a number of tetrasubstituted [2.2]paracyclophanes of type *23*; *23a* remains unchanged on heating for several hours at 250 °C. This thermal stability is compatible with the compound having an anti conformation.

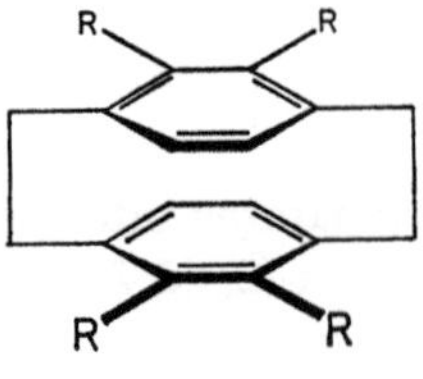

23a: R = $COOCH_3$
23b: R = CH_2OH
23c: R = COOH
23d: R-R = CO-O-CO

Some multi-substituted [2.2]paracyclophanes having additional methylene bridges present in the individual benzene moieties have been described by Nakazaki *et al.*[53]. The ^1H—NMR spectrum of *25a* shows an absorption at $\tau = 10.55$ (4 protons) which is assigned to the methylene-bridge protons of the aromatic nuclei, whereas *25b* shows no signal at similarly high field strength because the additional methylene groups apparently lead to increased conformational dynamic activity. The fact that *25a* does not form a π complex can also be ascribed to spatial shielding of both benzene nuclei. The [2.2]paracyclophane analogs *24a* and *24b* are unstable and have not been studied in detail.

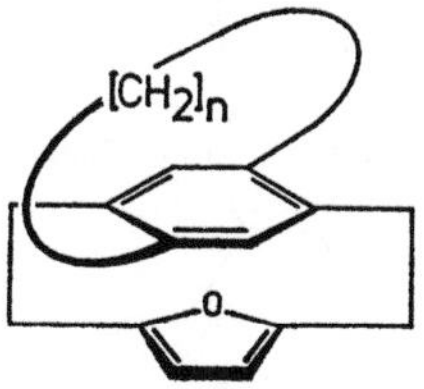

24a: n = 8
24b: n = 10

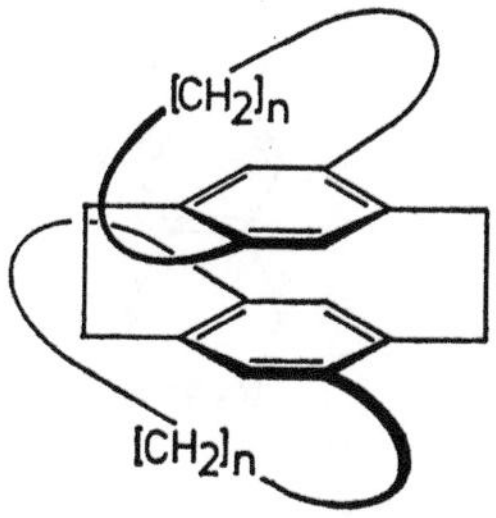

25a: n = 8
25b: n = 10

The UV spectra of 4,5,7,8-tetrafluoro[2.2]paracyclophane (*26*) [18] and of the octafluoro compound *27* [54] reveal the close relationship of these compounds to unsubstituted [2.2]paracyclophane (*2*). The absorption bands occurring between 286 and 291 nm, like those in the spectra of *2* can be attributed to deformation away from planarity of the aromatic rings. Compared with the fluorine-substituted open-chain analogs, these absorption bands are likewise bathochromically shifted by about 25 nm.

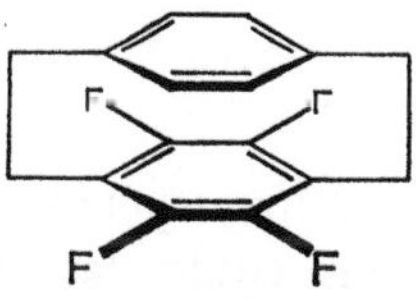

26

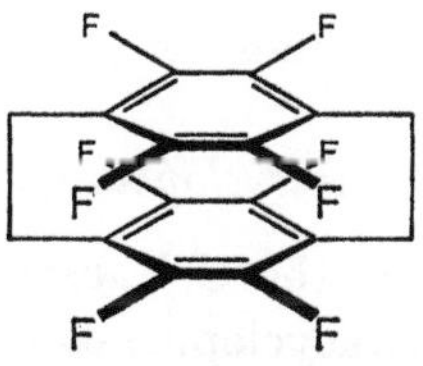

27

When the absorption spectra of *2*, *26* and *27* in the region around 300 nm are compared *26* shows a band at 297 nm, whereas *2* has a less intense shoulder at 302 nm. *27* lacks an absorption band in this region, which suggests that the 302-nm band of *26* is associated with an electron donor–acceptor transition, *i.e.* a π-π transannular interaction in an electron-withdrawing "π-acidic"[39] tetrafluorophenylene ring.

The proton resonance of *26* is also indicative of transannular interactions: interspatial transannular coupling of each aromatic proton with all four fluorine nuclei of the other ring induces splitting of the aromatic protons (quintuplet: $\tau = 3.18$, $J_{HF} = 0.8$ Hz). An alternative "long-range" coupling through the seven bonds is considered unlikely by Filler and Choe [18] since there is no evidence of HF coupling in the open-chain analog, 2,3,5,6-tetrafluoro-4,4'-dimethyl-dibenzyl (*28*) [55].

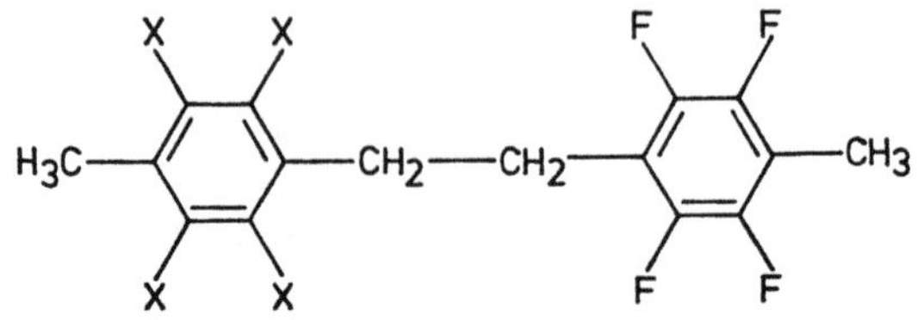

28: X = H   *29*: X = F

The bathochromic shift and hypochromism of the longest-wavelength absorption of 4,5,7,8,12,13,15,16-octafluoro[2.2]paracyclophane (*27*) relative to [2.2]paracyclophane (see Table 2) is attributed to both transannular fluorine–fluorine repulsion and increased deformation of the aromatic rings as a result of the larger spatial requirements of the fluorine atoms [56]. The longest-wavelength absorption of the analogous unbridged 2,3,5,6,2',3',5',6'-octafluoro-4,4'-dimethyldibenzyl (*29*) occurs at a much shorter wavelength [54].

The ^1H–NMR spectrum of *27* consists of a single complex CH_2 multiplet ($\tau = 6.71$). The shift of 0.25 ppm as against the corresponding [2.2]paracyclophane absorption [$\tau(CH_2) = 6.96$] is due to less effective shielding by the C_6F_4 system than by the C_6H_4 group. In the analogous open-chain system *29*, the benzyl protons absorb at lower field strength ($\tau = 6.97$) than those of the 4,4'-dimethyldibenzyl ($\tau = 7.27$).

2.2.2.4. *[2.2]Paracyclophanequinone*

By comparing the UV absorptionbands of [2.2]paracyclophanequinone (*30*), [4.4]paracyclophanequinone (*31*) [39], an equimolar mixture of 2,5-dimethylbenzoquinone and p-xylene, 2,5-dimethylbenzoquinone, and [8]paracyclophane, Cram and Day [57] were able to assign to a charge-tranfer transition the band at 340 nm found only with the intramolecularly complex molecule *30*.

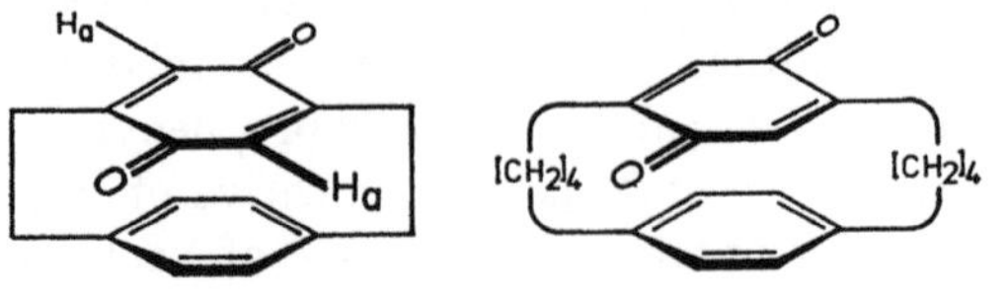

30   *31*

In the ^1H–NMR spectrum of *30* the H_a protons appear at considerably higher field strength ($\tau = 4.22$) than those of the hydrogen nuclei of 2,5-di-tert-butylbenzoquinone ($\tau = 3.53$).

2.3. [2.2]Paracyclophane-1-ene and [2.2]Paracyclophane-1,9-diene

The UV spectra of [2.2]paracyclophane-1-ene (*32*) and [2.2]paracyclophane-1,9-diene (*33*) [33] resemble that of [2.2]paracyclophane (*2*) (see Fig. 5); however, the absorption spectra of these olefins progressively overlap, thus reducing their information content.

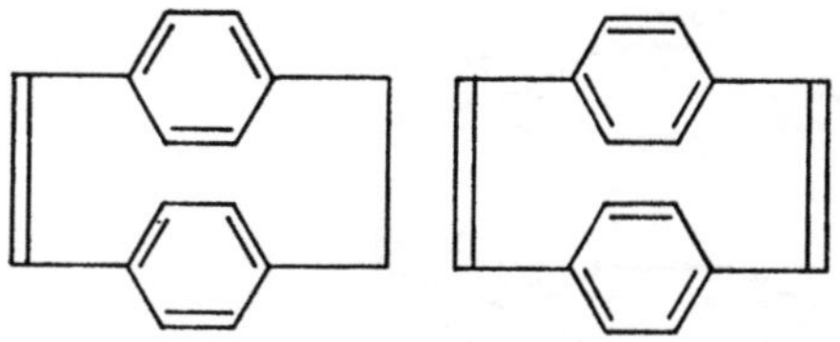

32 *33*

The unsaturated cyclic compounds *32* and *33* show none of the absorption characteristics of the open-chain analog, cis-stilbene, since the π electrons of the benzene rings and the C=C bonds cannot overlap because of the rigid architecture of the molecules.

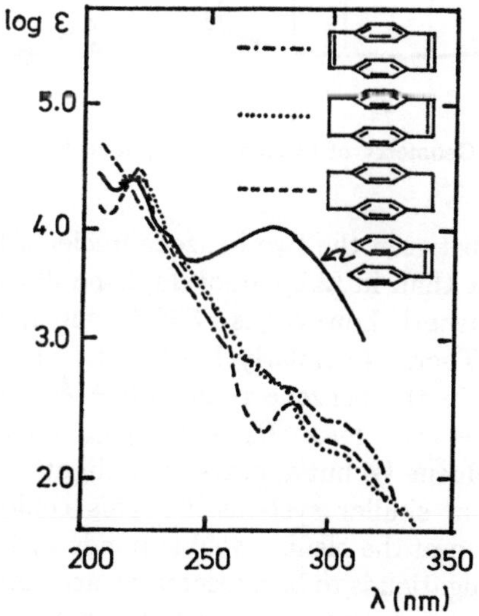

Fig. 5. UV spectra of *2*, *32*, *33*, and cis-stilbene

[2.2]Paracyclophane-1,9-diene (*33*) is undoubtedly the most highly strained representative of this series of bridged aromatic compounds. X-ray structural analysis [58] shows that the benzene rings of the molecule are puckered in a tublike fashion, and that the bridgehead atoms lie 0.166 Å and 0.178 Å outside the plane of the four unsubstituted aromatic carbon atoms. In [2.2]paracyclophane, the corresponding value is significantly smaller (0.13 Å). When viewed in the end-on position, the benzene rings in a molecule of *33* appear to be folded about the axis formed by the bridgehead atoms (Fig. 6c).

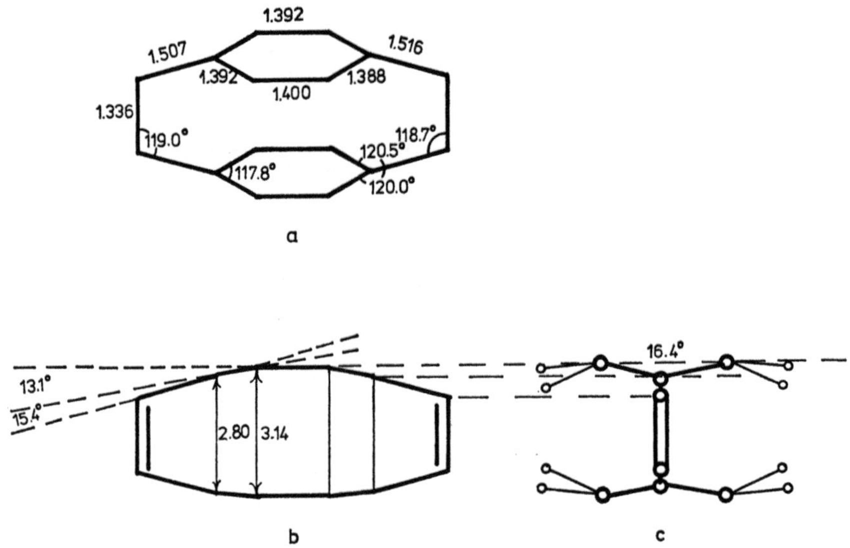

Fig. 6. Geometry of [2.2]paracyclophane-1,9-diene (*33*)

The distances between the two benzene nuclei (2.80 and 3.14 Å) are significantly longer than in [2.2]paracyclophane (2.75 and 3.09 Å). The olefinic C=C bond length, however, is 1.336 Å compared to 1.334—1.337 Å in other alkenes. There is similarly hardly any change in the average C—C bond length in the benzene rings (1.399 Å) compared to 1.393—1.397 Å in benzene. The mean C_{aryl}—$C_{methine}$ bond length (1.512 Å) is the same as in toluene [59] but appears to be slightly stretched compared with such bonds in similar systems [60]. This could be attributed to reduced conjugation of the olefinic double bonds with the benzene rings. A decrease in conjugation is to be expected on account of the orthogonality of the π orbitals of the two unsaturated structural elements; striking evidence of this is found in the UV spectrum of the diene (see above).

Gantzel and Trueblood [11] estimated the total strain of the diene as 39 kcal/mol. The ring strain is by no means distributed over all bond lengths and angles; it is mainly taken up by deformation of the benzene rings and to a lesser extent by extension of the C_{aryl}-$C_{methine}$ bond lengths and deformation of the dihedral angle.

2.4. Analogs of [2.2]Paracyclophane

Two interconvertible diastereoisomers *A* and *B* of [2.2](1,4)naphthalenophane (*34*) [16,61,62] are known. The H_a and H_b proton resonances lie at higher field strength in the syn form *B* than in the anti form *A*

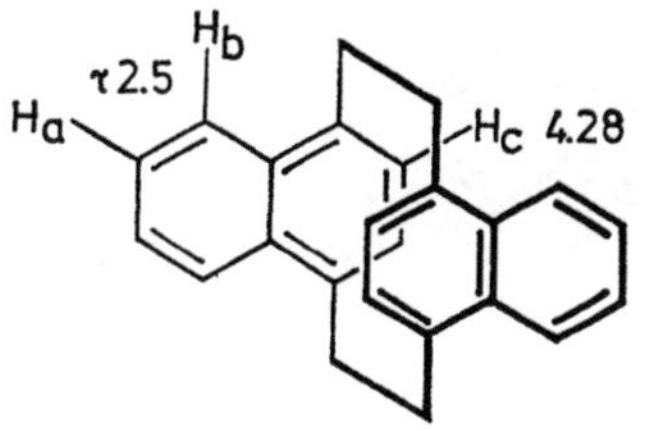

anti-*34*(*A*)

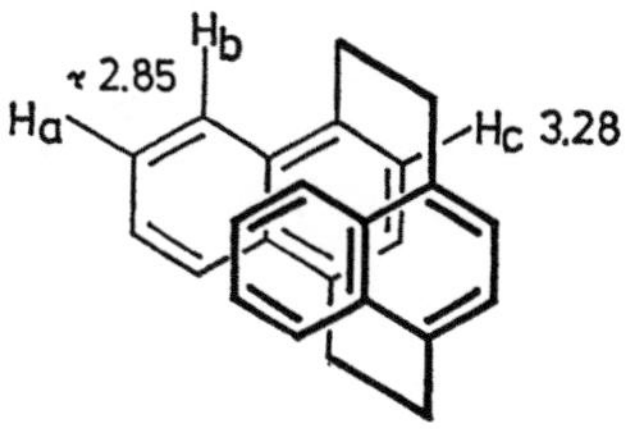

syn-*34*(*B*)

whose spectrum closely resembles that of 1,4-dimethylnaphthalene. On the other hand, the H_c protons of the anti form absorb at higher field strength, while the syn H_c protons are less shielded.

Characteristic shifts due to paramagnetic ring currents are also met with in the singly annellated [2]paracyclo[2](1,4)naphthalenophane (*35*) [16].

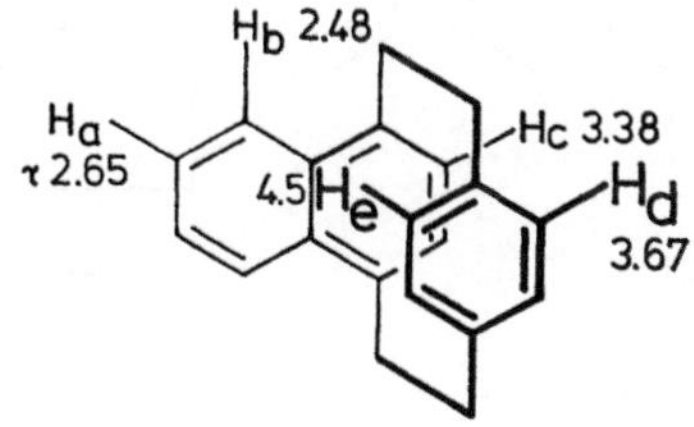

35

The ¹H—NMR spectra of the recently prepared [2.2](1,4)anthracenophanes *36* [63] resemble those of the isomeric naphthalenophanes *34*. In the anti conformer *A*, the H_d proton, which lies directly above an anthracene ring, absorbs at much higher field strength than the syn H_d proton.

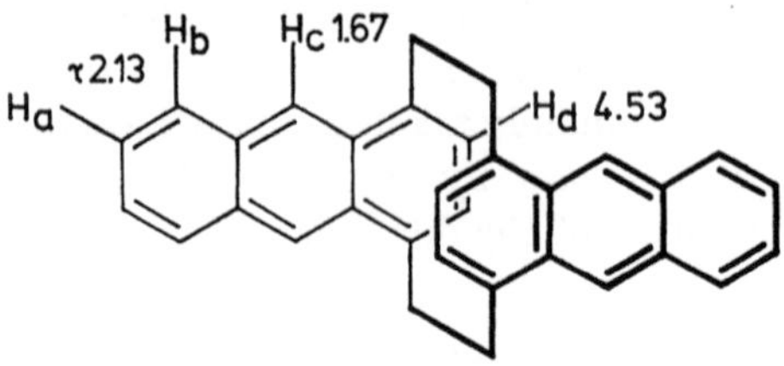

anti-*36*(*A*)

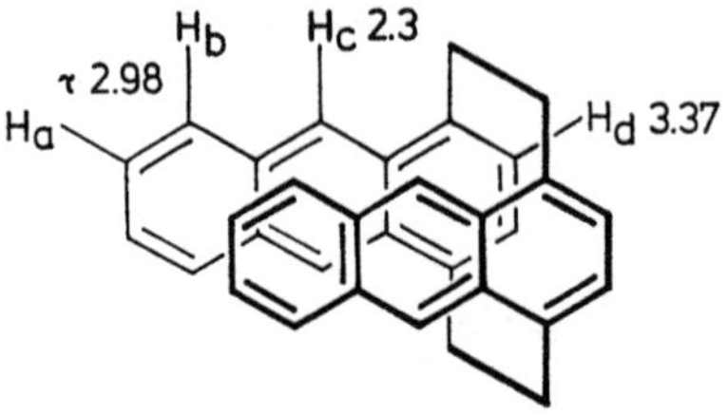

syn-*36*(*B*)

Cram *et al.*[16] have synthesized a number of compounds *37*–*40* having structures analogous to [2.2]paracyclophane. Although compounds *38*–*40* could not be isolated in pure form, assignment of the ^{1}H–NMR absorptions presented little difficulty.

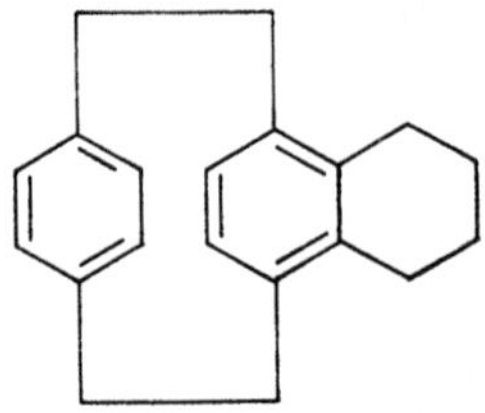

37

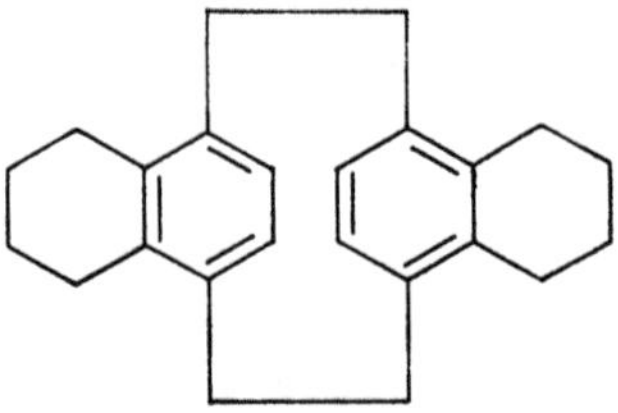

38

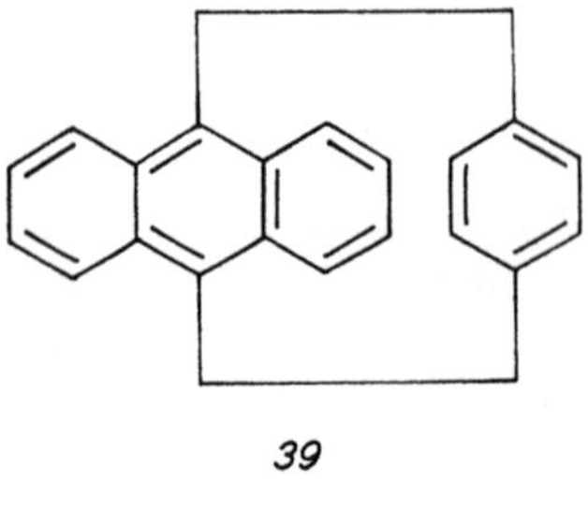

39

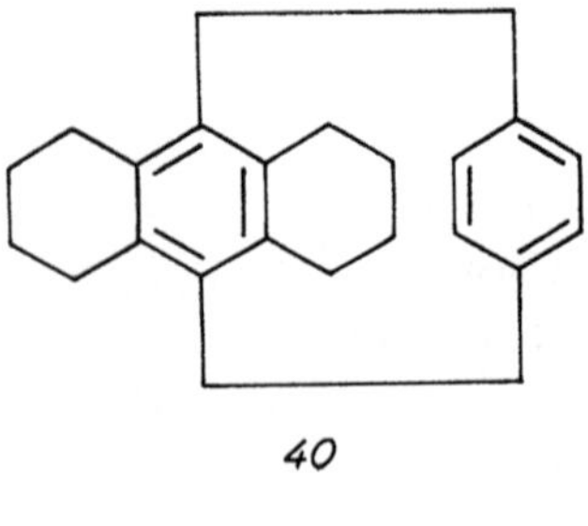

40

The following compounds which have a benzene moiety of [2.2]paracyclophane replaced by a heteroring (*e.g.* furan) may be counted among the analogs of [2.2]paracyclophane, as their stereochemistry has many features in common with that of the carbophanes [64] *34*—*36*.

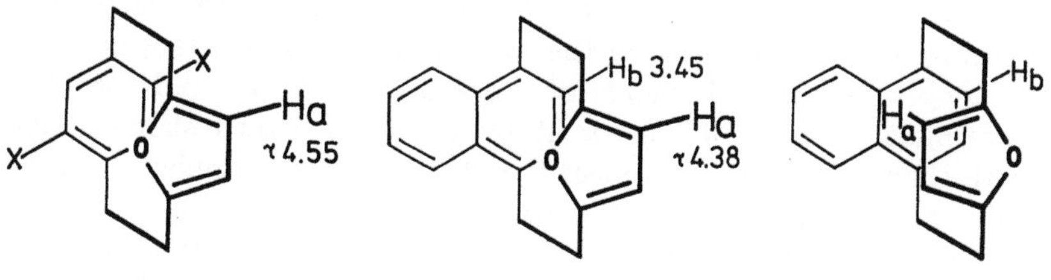

41a: X = H
41b: X = D

anti-*42*

syn-*42*

Assignment of an anti configuration to a [2](2,5)furano[2](1,4)naphthalenophane (*42*) synthesized by Wasserman and Keehn [65] followed from a comparison of its ^{1}H—NMR spectrum with that of [2]paracyclo-[2](2,5)furanophane (*41a*) [66]. The absorption band assigned to the β-furanoid proton H_a in the spectrum of *42* ($\tau = 4.38$) appears in the same region as the corresponding band for *41a*. In the case of syn-*42*, a chemical shift would be expected due to the transannular shielding effect of the naphthalene nucleus.

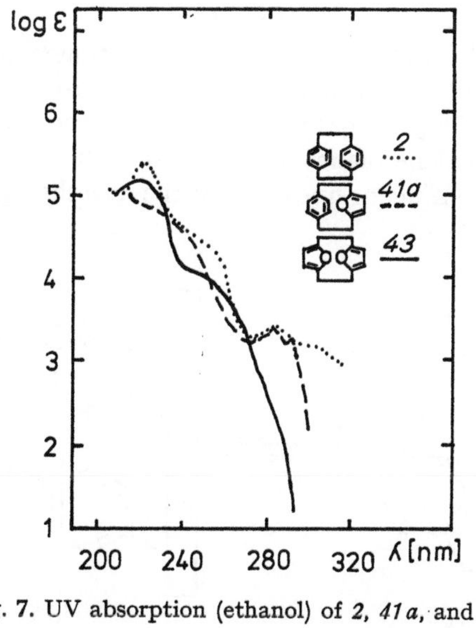

Fig. 7. UV absorption (ethanol) of *2*, *41a*, and *43*

The ^{1}H–NMR spectrum of *41 b* is temperature-dependent. According to dynamic NMR spectroscopical studies the underlying diastereotopomerization [67a)] involves an Arrhenius energy of 11.1 ±0.3 kcal/mol [67b)].

The UV absorption curves of compounds *2*, *41 a*, and *43* are reproduced in Fig. 7 [66b)]. The spectrum of *41 a* can to all intents and purposes be described as a superposition of the spectra of *43* and [8]paracyclophane. The longest-wavelength bands of *2* are absent in *41 a* and the other transannular bands of *41 a* occur at about 244 nm, a much shorter wavelength than in *2*. It would appear therefore that transannular electronic effects in the excited state of *41 a* are less significant than in *2*.

[2.2](9,10)Anthracenophane (*44*), first described by Golden [68)], is an orange-colored compound that crystallizes monoclinically.

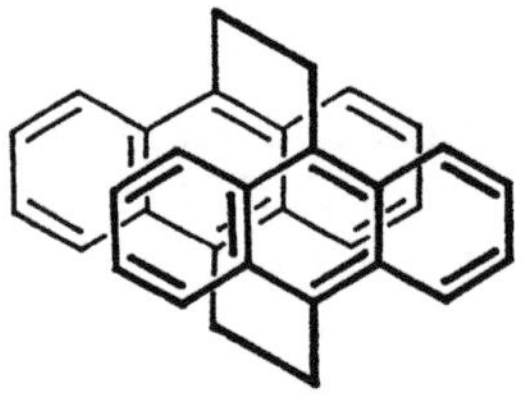

44

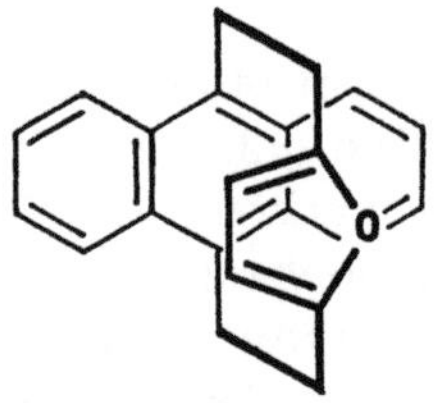

45

The photochemical behaviour of this compound [69)] and the reactions of the analogous [2.2]paracyclophane [2](9,10)anthraceno[2](2,5)furanophane *45* [70)] are described in Section 3.2.

The ^{1}H–NMR spectrum of [2.2](1,6)naphthalenophane (*46*) is not consistent with a symmetrical structure *46b* [71)]. According to molecular models, the CH_2–CH_2 bridges in *46a* are not perpendicular but inclined toward the naphthalene planes and skewed with respect to each other. Rotation of the naphthalene nuclei about the axis through the bridgehead atoms can be ruled out.

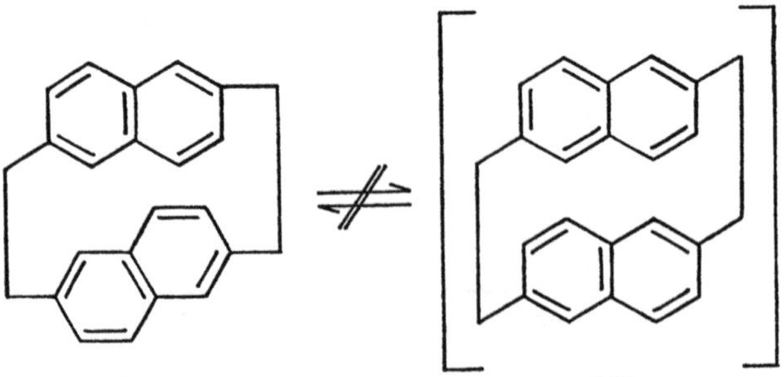

46a *46b*

The UV spectrum of the chiral molecule *46a* differs from 2,6-dimethylnaphthalene by a long-wave shift of the first absorption band and almost complete disappearance of the sharp vibration structure of the naphthalene spectrum.

The shift of the aromatic protons of [2.2](4,4')biphenylophane (*47*) [71] to higher field strength is smaller than in [2.2]paracyclophane ($\tau(H)_{arom} = 3.63$ in $CDCl_3$). Furthermore, the spectra of *47* and 4,4'-dimethylbiphenyl differ less from each other than do the spectra of [2.2]paracyclophane and p-xylene.

There is obviously less perturbation of the π-electron system in *47* than in [2.2]paracyclophane, since the deformation of the valency angles is distributed over a greater number of bonds.

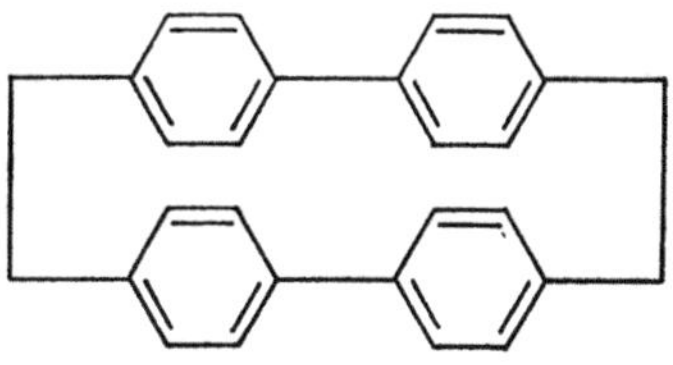

47

The ^1H–NMR spectrum of 5,6,17,18-tetrahydro[2.2](2,7)phenanthrenopane (*48*) is compatible with the dihydrophenanthrene units being in an anti position. Dehydrogenation of *48* with 2,3-dichloro-5,6-dicyano-p-benzoquinone in benzene gave [2.2](2,7)phenanthrenophane which, owing to the similarity of its ^1H–NMR spectrum to that of 2,7-dimethylphenanthrene, is assumed to have its phenanthrene units in the anti position [71].

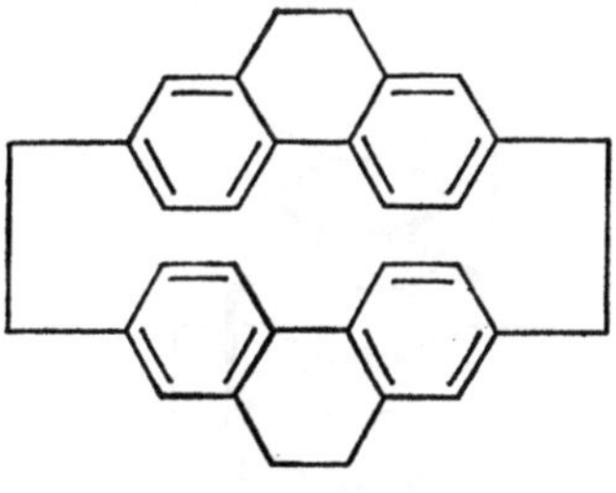

48

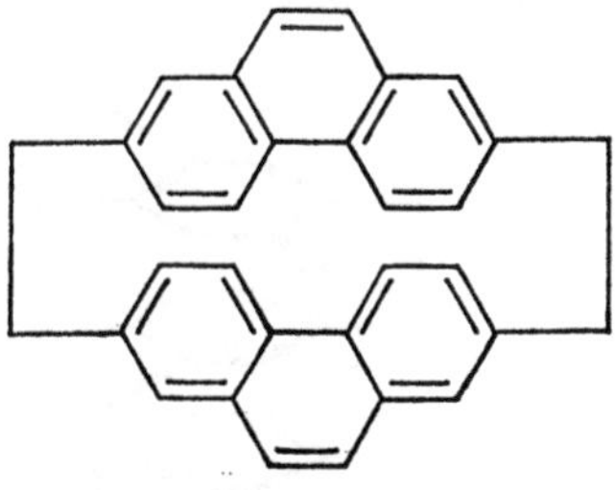

49

Bruhin and Jenny [72] recently prepared the [2.2](2,5)pyridinophanes *50a*–*50d* by Hofmann elimination; the compounds were separated by gas chromatography. Their ^1H–NMR spectra provide clear structural

Dynamic NMR spectroscopic measurements indicate, that the free enthalpies of activation of *57* and *59* are 10.2 kcal/mol and 11.9 kcal/mol, respectively. Similar conformational behavior is exhibited by the [2.2](2,5)heterophanes *43*, *61*, and *62*. The thiophene-containing heterophanes *61* and *62* [81,82] are conformationally rigid up to 200 °C and are not subject to ring inversion, whereas the furanophane *43* is conforma-

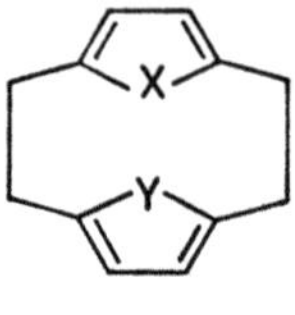

61: X = Y = S
62: X = S; Y = O
43: X = Y = O

tionally mobile ($\Delta G_c^{\ddagger} = 16.8$ kcal/mol) [82]. In view of the entirely different geometry and ring strain of five-membered rings, no direct comparison of the inversion barriers of *57*, *59*, and *43* can be attempted until additional data are available on the entropy and enthalpy of activation.

3. Molecular Geometry and Chemical Properties

3.1. Transannular Interactions in Chemical Reactions

In addition to chemical reactions which result in direct valency coupling, other reactions, for example, those that lead to the formation of charge-transfer complexes are rather generally characteristic of the [2.2]paracyclophane system.

3.1.1. Reactions at the Aromatic Nucleus

3.1.1.1. π-Complex Formation

Cram *et al.* have investigated whether and how the electronic effects of the substituents in one ring are transferred to the second ring of [2.2]paracyclophane.

A number of tetracyanoethylene (TCNE) π complexes of monosubstituted [2.2]paracyclophanes were examined both spectroscopically and kinetically. The position of λ_{max} of the longest-wavelength charge-transfer band in the UV spectrum of the complexes, and in some cases estimated equilibrium constants [76] were used as a measure of the relative π-base strengths of the substituted cycles [15]. From the results of this investigation it was at once clear that complexes of tetracyanoethylene and [2.2]paracyclophane with electron-releasing substituents (Type *A*,

Fig. 9) are more stable than those with electron–withdrawing groups (Type *B*).

The authors assume that the electron–releasing character of the substituent group in complexes of type *A* (see below) must make the substituted ring the more basic one, exhibiting more π–π interaction with the TCNE molecule, whereas in complexes of type *B* the electron–withdrawing character of the substituents probably means that they deactivate the ring to which they are attached. This assumption is supported by the finding that the complex with 4-acetyl[2.2]paracyclophane is even more stable than that with p-xylene, despite the electron–withdrawing character of the substituents.

The competing effects of the electron–releasing ability of the non–complexing benzene ring against the electron–withdrawing character of the substituents (acetyl, nitrile) are illustrated in Fig. 9.

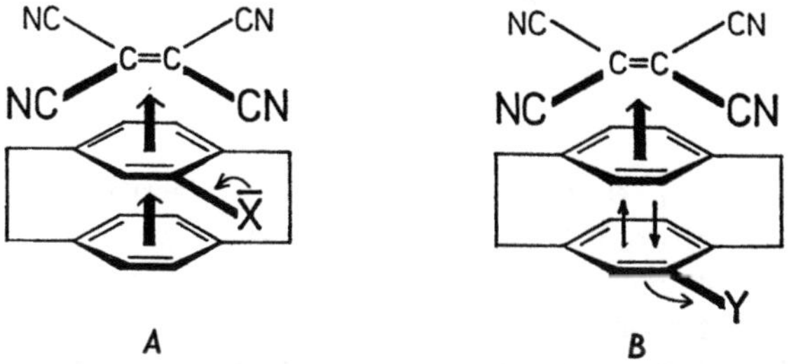

Fig. 9. π–π Complexes with electron–releasing substituents X (*A*) and with electron-withdrawing substituents Y (*B*)

On the basis of these considerations, the authors suggest that a roughly linear relationship might exist between the transition energies (E_t) for the charge–transfer band and Hammet–type substituent constants, σ_m.

3.1.1.2. Transannular Directing Effects in Electrophilic Substitution

Chemical reactions have also provided valuable information regarding transannular electronic and steric effects.

Depending upon substituents, transannular interactions in the [2.2]paracyclophane system are characterized by the steric or electronic effects of one aromatic nucleus on the physicochemical behavior of the other aromatic ring. The transannular reactions themselves, of course, are very dependent upon molecular geometry.

The relative rates of acetylation in competition experiments in the [m.n]paracyclophane series [38] may be interpreted in terms of transannular electronic and steric effects. If the rate of acetylation of [6.6]paracyclophane [(*1*), $m = n = 6$] is is taken as one, the relative acetylation rates of the [4.4]–, [4.3]–, and [2.2]paracyclophanes are 1.6, 11, and >48, respectively. As the aromatic rings come closer together, the rate of entry of the first acetyl group into the nucleus increases, while that of the second acetyl group decreases. Both effects clearly indicate that the positive partial charge can be distributed over both benzene rings in the monoacetylation transition state (*64*).

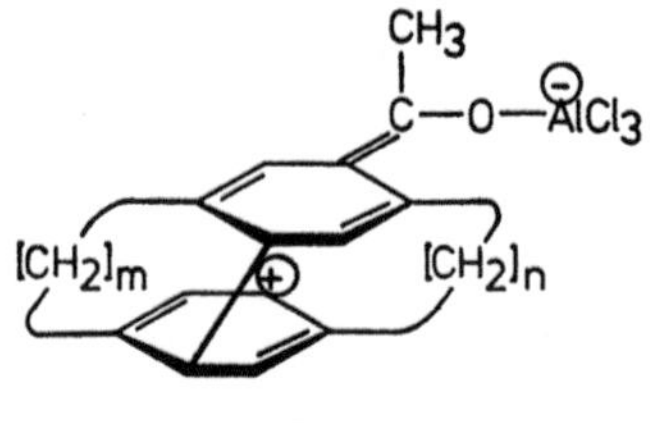

64

Reich and Cram [83] studied the patterns of electrophilic substitution of the monosubstituted [2.2]paracyclophanes. It was at once clear that the directive influences of the substituents X (see below) could not be correlated with transannular resonance effects in the ground state [84]. The product pattern predicted on the basis of electrostatic ground–state models, such as the canonical structures *65* for electron–releasing and *66*

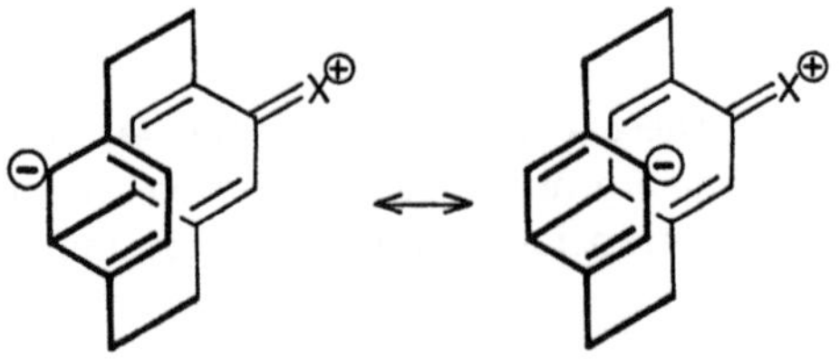

65(X electron-repulsing)

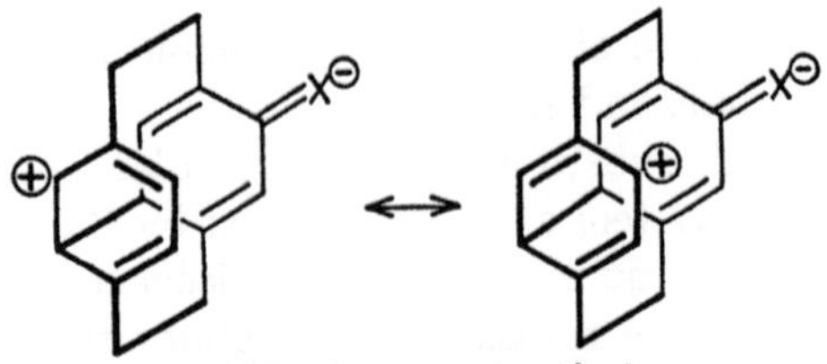

66(X electron-attracting)

for electron-withdrawing substituents, does not correspond with the observed results of secondary substitution.

The resonance structures *65* suggest that the pseudo-meta and pseudo-geminal positions are the preferred sites for substitution; however, it has been observed (*e.g.* in the bromination and acetylation of 4-bromo[2.2]paracyclophane (*67*)) that there is predominant formation of pseudo-ortho and pseudo-para products [83b,85] (see Table 3):

Table 3. Product distribution in the second electrophilic substitution of mono-substituted [2.2]paracyclophanes

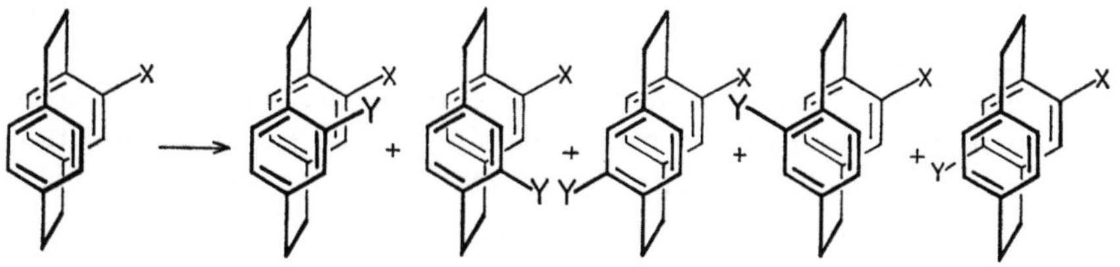

X	Y	% pseudo-gem.	pseudo-ortho	pseudo-para	pseudo-meta	para
67: Br	*68*: Br		16	26	6	5
67: Br	*69*: $COCH_3$	<0.7	6(9) [1]	41(58) [1]	(<12) [1]	17(11–23) [1]
70: CO_2CH_3	*71*: Br	89				
72: $COCH_3$	*73*: Br	59				
74: NO_2	*75*: Br	70	3	6	8	
76: CN	*77*: Br		16	28	26	

[1] Yields determined indirectly [83b].

The canonical structures such as *66* suggest that the pseudo-ortho and pseudo-para positions would be the preferred site(s) for electrophilic substitution. However, the bromination of 4-carbomethoxy- and of 4-acetyl[2.2]paracyclophane gave only the pseudo-geminal substituted product [83b].

From the experimental results it can be assumed that in the [2.2]-paracyclophane system substitution predominantly occurs pseudo-geminally to the positions of the most basic prior substituents in the ring. In the bromo compound, for example, the most basic positions of the substituted ring are the ortho and para positions (see Fig. 10). Bromination and acetylation actually do occur predominantly in the pseudo-ortho and pseudo-para positions.

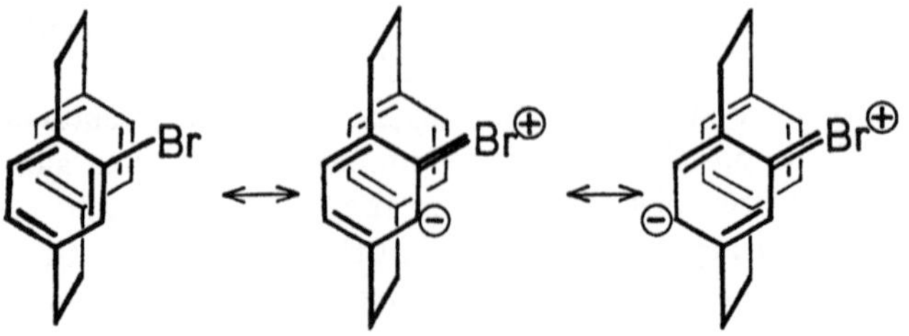

Fig. 10 Canonical structures of 4-bromo[2.2]paracyclophane (*67*) [83b)]. Cram called the polar formulae "intraannular resonance structures"

In the nitro- and carbonyl-substituted [2.2]paracyclophanes the most basic position is at the oxygen of the substituent, and substitution occurs in the pseudo-geminal position.

Correlation of transannular directive properties with the basicity of the positions and groups pseudo-geminal to the site of substitution suggests that the observed directive effects involve the participation of a neighboring internal base in the product-determining step of the substitution reaction. According to Cram, the already substituted aromatic nucleus can act in this way. The mechanism of the electrophilic substitution leading to pseudo-para substitution is discussed, taking as examples 4-bromo[2.2]paracyclophane (*67*) and the monodeutero compound (*78*) (see Fig. 11).

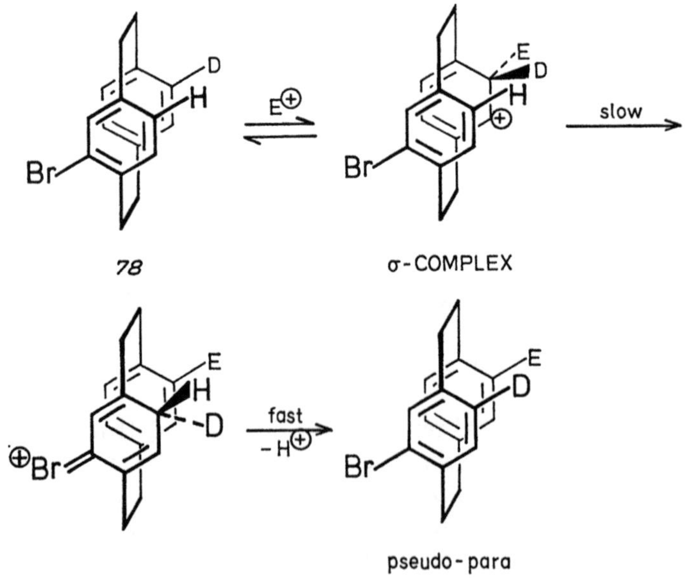

Fig. 11. Mechanism of the electrophilic substitution leading to pseudo-para substitution

The electrophile E^+ attacks the unhindered side of the still unsubstituted second aromatic ring. A proton (deuteron) is transferred from this ring to the second, originally substituted ring, from which it leaves the molecule. Thus, the electrophile enters, and the proton (deuteron) leaves the [2.2]paracyclophane system by the least hindered paths. Some migration of deuterium could be detected in the bromination of 4-methyl[2.2]paracyclophane (*79*). The proposed mechanism is supported by the kinetic isotope effects (k_H/k_D) found for bromination of p-protio and p-deuterio-4-methyl[2.2]paracyclophanes in various solvents; these isotope effects demonstrate that proton loss from the σ complex is the slowest step.

The most striking example of transannular directive effects is found in the exclusively or predominantly pseudo–geminal substitution observed in the bromination of [2.2]paracyclophanes with substituents containing basic oxygen (carboxy, methoxy, acetyl, nitro). The oxygen atom of the functional group is obviously in an ideal position for accepting a proton from the pseudo–geminal position of the σ complex. A mechanism differing from that mentioned above is formulated for the bromination of the ester *70* (Fig. 12). Here the oxygen atom of the electron–with-

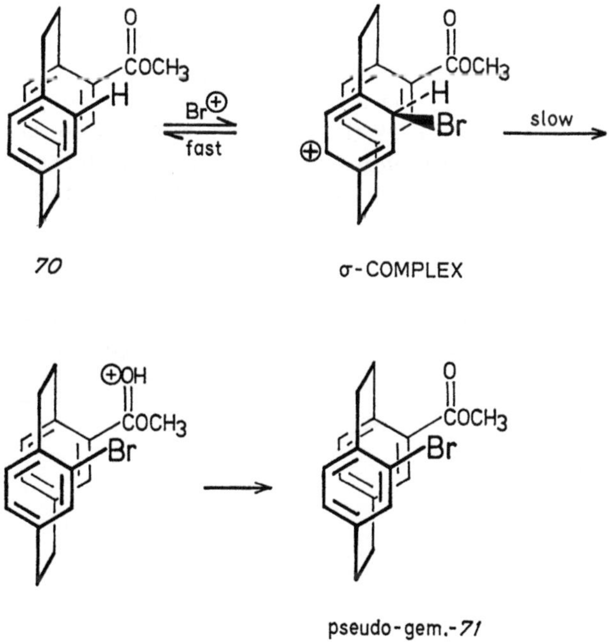

Fig. 12. Mechanism of the electrophilic substitution leading to pseudo–geminal substitution

drawing primary substituent acts as the intramolecular base and assists proton removal. The geometry of the cyano group, on the other hand, precludes its acting as the neighboring group for transannular removal of a proton; no pseudo-geminal disubstitution product is found in the bromination of *76* [86].

3.1.1.3. Transannular Reactions

The unique molecular geometry of the [2.2]paracyclophane system is demonstrated by the ability of the [4.4]paracyclophane *80a* ($n=4$) to undergo transannular ring closure *80*→*81* [37], whereas no ring closure occurs when $n=2$ [21].

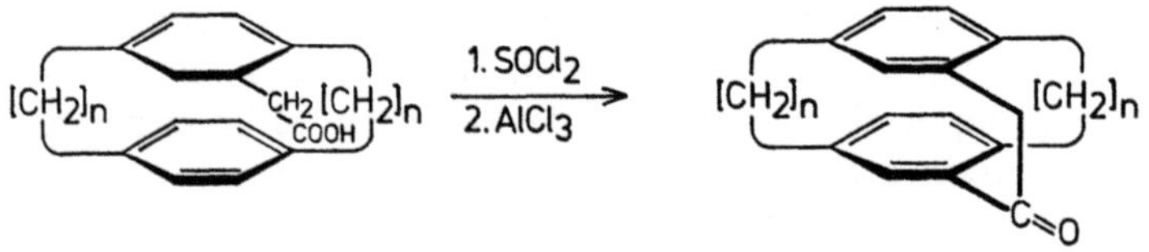

80a: *n* = 4
80b: *n* = 2

81a: *n* = 4
81b: *n* = 2

The occurrence of an additional two-membered bridge as intermediate in a [2.2]paracyclophane system has been postulated by Forrester and Ramasseul [47b]. During the synthesis of the bishydroxylamine *82*, a precursor of the diradical *19* these authors were able to isolate as by-product the pseudo-para hydroxy compound *83* which is converted into the violet quinone *84* through oxidation by atmospheric oxygen or reac-

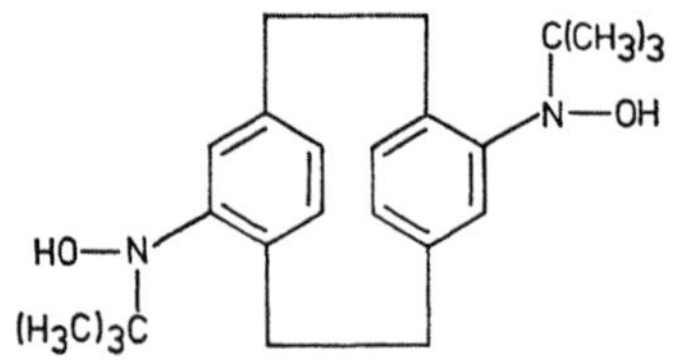

82

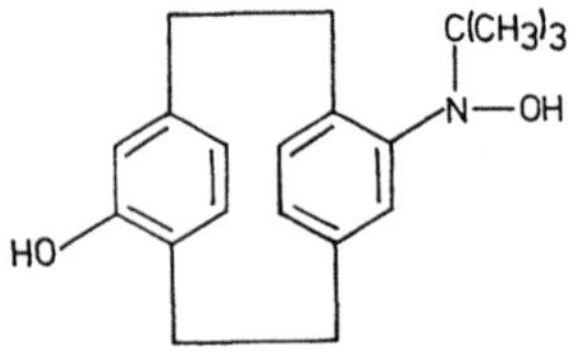

83

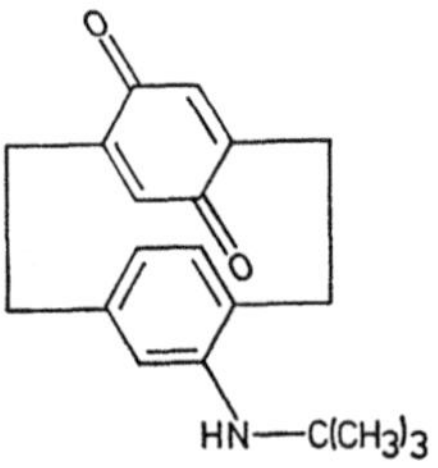

84

tion with silver oxide. The following reaction mechanism is proposed:

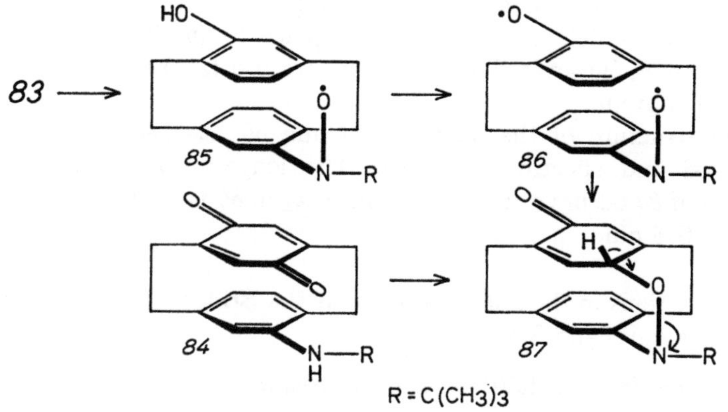

The decisive step in this reaction is the transannular migration of the N-oxide oxygen atom (*86*→*87*). Such an intramolecular attack seems quite plausible, since model studies demonstrate that the oxygen atom of the N—O group in *86* can approach within 2.6 Å of the pseudo-geminal position in the other benzene ring. Analogous intermolecular reactions, such as oxidation of phenols to quinones by N-oxides, are known [87].

The black orthoquinone *89* appears to be formed in the same way from the pseudo-ortho disubstituted [2.2]paracyclophane *88*. On the other hand, the corresponding oxidation of the pseudo-meta disubstituted paracyclophane *90* yielded only an incompletely identifiable mixture of compounds which included neither *84* or *89*. This finding supports

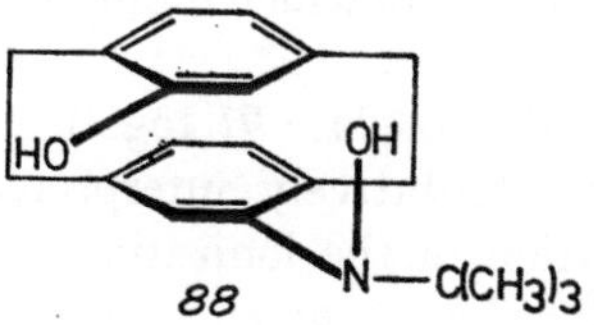

88

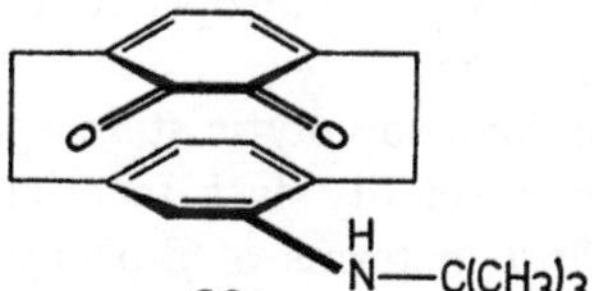

89

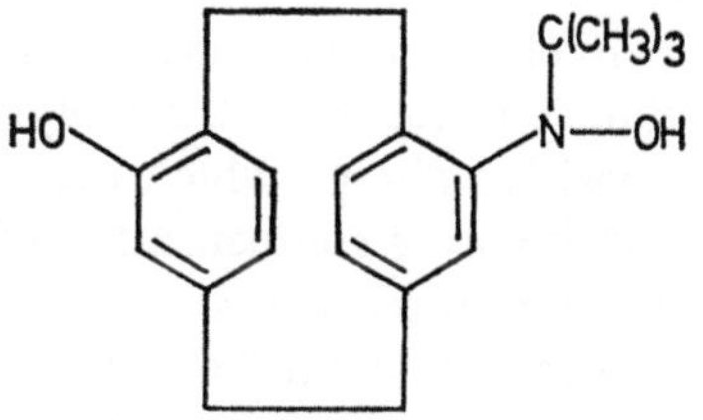

90

the hypothesis of an intramolecular oxygen transfer process. Such a process is improbable for the pseudo–meta substituted compound *90,* since intermediate coupling in the pseudo–ortho or pseudo–para position to the NO group must take place to produce an ortho– or para–quinone; however, both positions are too far away for attack by the oxygen atom.

The structure of the quinones *84* and *89* has been confirmed by analytical and spectroscopic data. The longest–wavelength UV absorption of *84* occurs at 545 nm, while λ_{max} in *89* is shifted bathochromically to 605 nm.

3.1.2. Reactions in the Side Chain and at the Bridge: Dynamics and Stereospecificity

3.1.2.1. Substitution, Addition, and Elimination

Cram *et al.*[88] have studied the stereochemical course and solvolysis reactions of optically active 1-tosyloxy[2.2]paracyclophane *91* (see scheme *91—95*). Reactions *92*→*91*, *92*→*93*, *92*→*94*, and *92*→*95* were all shown to proceed as expected, with complete retention of configuration.

In contrast, the results obtained in the methanolysis, acetolysis, and trifluoroacetolysis of the tosylate *91* were not the expected ones. Cram obtained the methyl ether *93*, the acetate *94* and the trifluoroacetate *95* with the same configuration and optical purity as in the direct synthesis from the alcohol *92*. These solvolyses at the bridge carbon atom of [2.2]paracyclophane therefore proceed with complete retention of configuration. The rate of acetolysis of the tosylate *91* also deviates considerably from that of aliphatic secondary tosylates; it is some 100 times faster than that of 2-butyl tosylate and about the same as that of α-phenylneopentyl tosylate, acetolysis of which is only slightly stereospecific.

The high solvolytic stereospecificity of the tosylate *91* together with the unexpectedly fast reaction rates was tentatively interpreted by Cram [88b] in terms of β-phenyl participation in the ionization step to produce a highly strained bridged carbonium ion *96* which is opened in a second reaction step to give the final product. Both the formation of *96* and its opening must involve complete inversion in order to ensure retention of stereospecificity in the overall solvolytic process.

The ion *96* carries a positive charge which can be distributed over both rings; formation of *96* involves compensation of bond angle strain, leading to facilitation of π–π repulsion strain between the neutral benzene nuclei in the starting compound. The twisting of the system in *96* corresponds to the somewhat twisted crystal structure of the [2.2]-paracyclophane molecule.

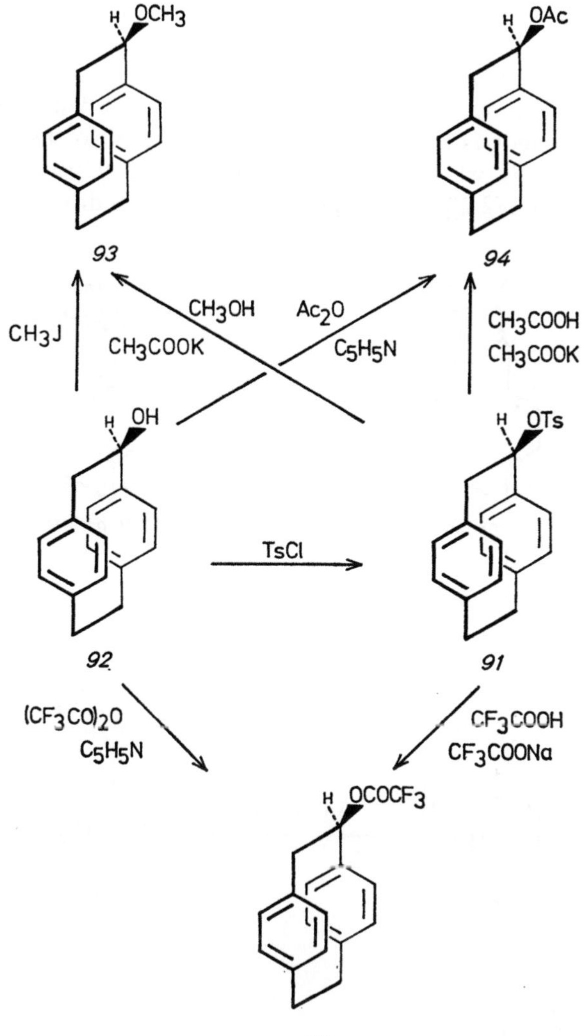
H OCH3
H OAc
93
94
CH3J
CH3OH
CH3COOK
Ac2O
C5H5N
CH3COOH
CH3COOK
H OH
H OTs
TsCl
92
91
(CF3CO)2O
C5H5N
CF3COOH
CF3COONa
H OCOCF3
95

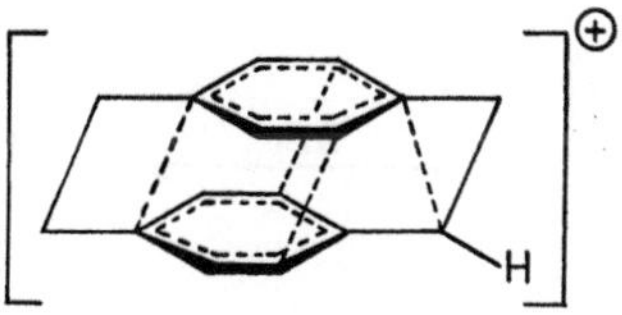
H

96

According to Cram [88a)], the stereochemical course of several polar addition and substitution reactions at the bridge position of the [2.2]-paracyclophane system is best explained on the basis of a species similar to *96*: Reaction of [2.2]paracyclophane-1-ene (*32*) with bromine or deuterium bromide is postulated to lead exclusively to the cis-addition products *97* and *99*, which retain configuration on acetolysis. The reaction of cis-1,2-dibromo[2.2]paracyclophane (cis- *97*) with lithium bromide in

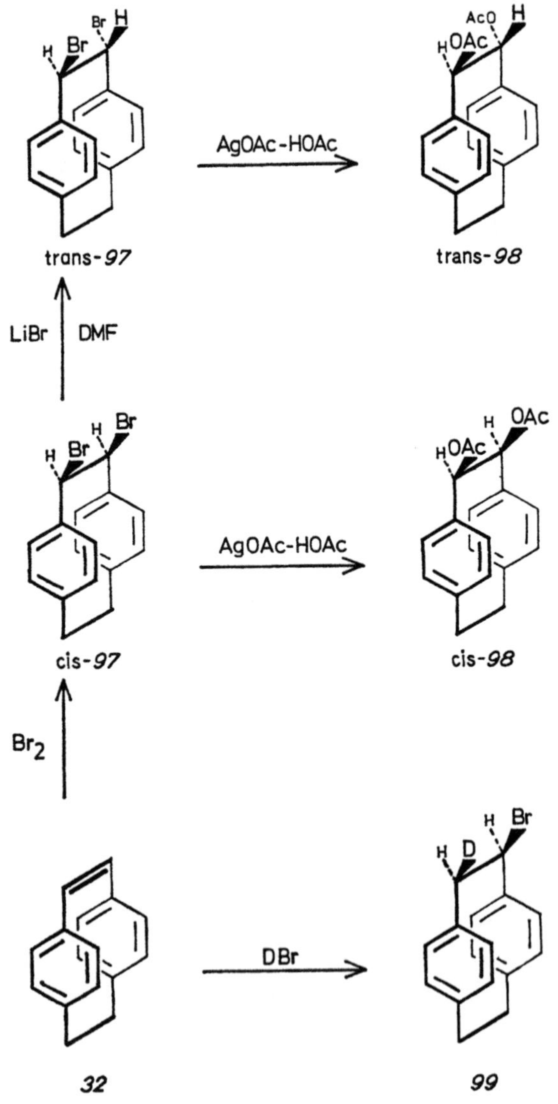

dimethylformamide leads exclusively to trans- *97*, which on acetolysis again affords only trans-diacetate.

3.1.2.2. [2.2]Paracyclophanyl as Neighboring Group

Cram and Singer [89] have investigated the ability of [2.2]paracyclophane to function as a neighboring group in carbonium ion-forming systems such as *100a* and *100b*.

Measurement of reaction rate and activation parameters showed that in solvolytic reactions the [2.2]paracyclophanyl system is a much more active neighboring group than is the phenyl nucleus in its open-chain counterparts *101a* and *101b*: Aryl participation (charge delocalization) is greater in paracyclophanyl bromide (*100a*) than in *101a*.

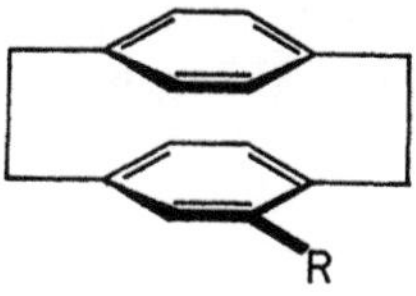

100

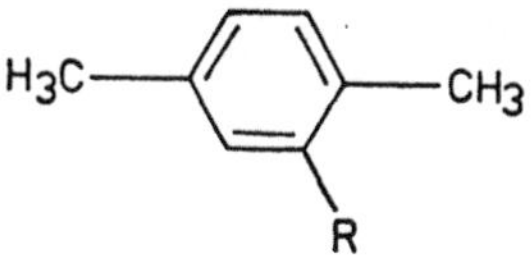

101

a: R = CH_2Br
b: R = CH_2-CH_2-OTs
c: R = CH_2-CD_2-OTs

Hydrolysis of optically pure bromide *100a* in dioxane/water gives the optically pure alcohol. This is consistent with the transannular p-xylylene ring participating in carbonium ion formation only through π–σ charge delocalization (π–σ resonance) of the type *104* rather than by direct participation in replacement of bromide via a transannularly bridged ion such as *102a*. In the latter case, racemization would be expected to take place:

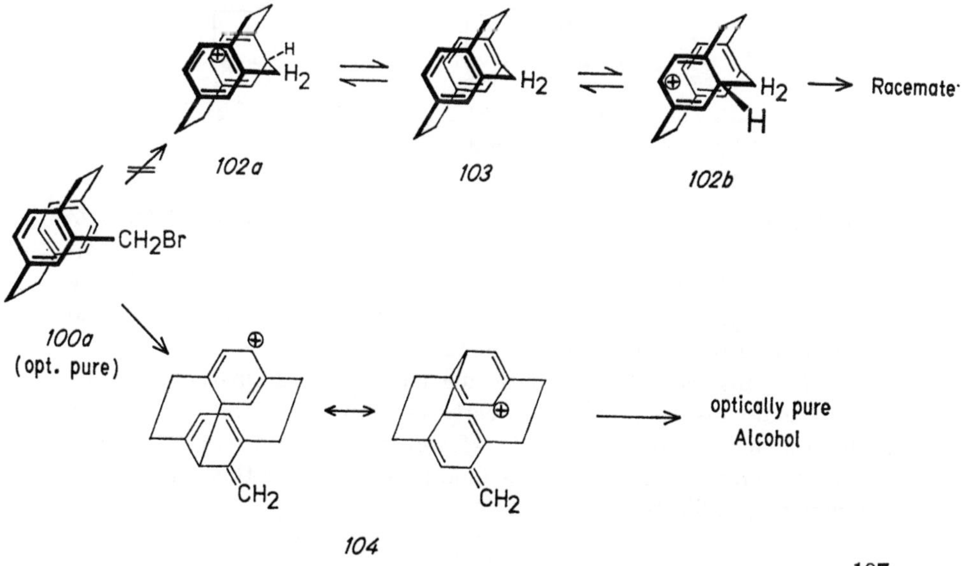

Cram and Singer also noted a remarkable difference in the solvolysis of the systems *100b* and *101b*. On acetolysis of the α–C deuterated model compound *101c* the deuterium was found to be scrambled with 46% D recovered in the β position (on hydrolysis 26% D was recovered the β position), whereas in the acetolysis and formolysis of *100c* the deuterium remained totally undisturbed in the α position.

According to Cram and Singer, there are two possible explanations for aryl participation at the ß–C atom of the paracyclophane system:

A. An unsymmetrical ethylenephenonium ion *105* represents a transition state which, on steric grounds, can be attacked by the solvent only at the α–C, while the deuterium originally at α–C remains fixed. The experimental evidence strongly supports the assumption that *105* occurs as a discrete intermediate, at least in the acetolysis and formolysis of *100c*, which would imply that a positive charge in one ring of the paracyclophane system is delocalized into the second ring through π–σ charge delocalization.

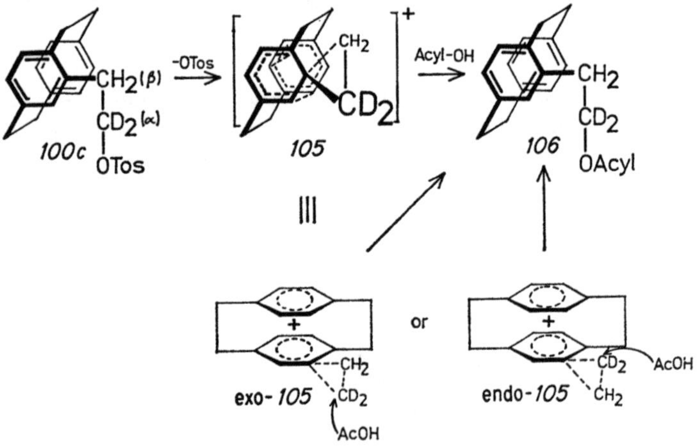

Recent studies on the solvolysis of the diastereomeric tosylates *107* support this hypothesis. The acetolysis and formolysis of these compounds proceed with high retention of configuration [90] while in ordinary simple acetolysis of tosyl esters of secondary alcohols containing no neighboring groups, inversion predominates.

B. The unsubstituted aromatic nucleus of the [2.2]paracyclophane system could also participate directly in ionization. This would give an intermediate product like *108a*. However, the absence of *109* and of α- and β-deuterated species in the final product makes this type of aryl participation unlikely.

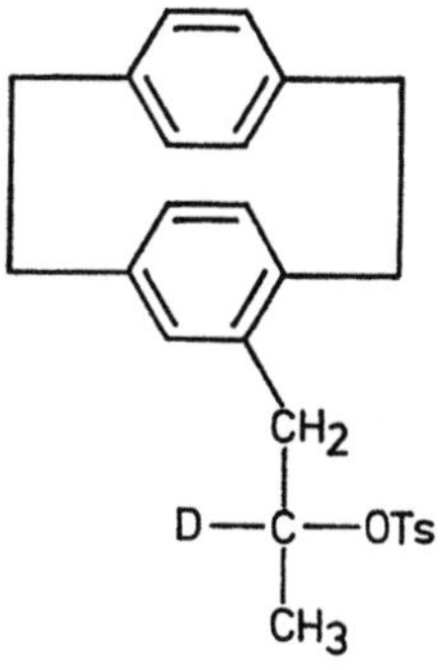

107

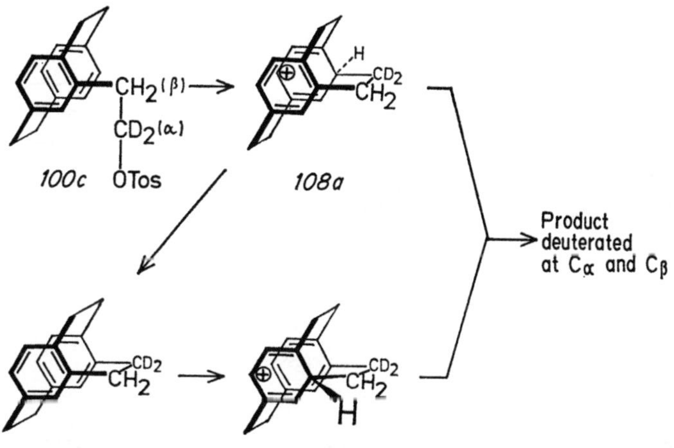

100c *108a* *109* *108b*

Nugent *et al.*[91,92] have studied the stereochemistry and dynamics of the paracyclophanylphenonium ion in some detail, particularly the question of whether the phenonium ion is exclusively formed from and neutralized at either the exo or endo methylene group, or CD_2 group. The observed rate of formolysis of 1-tosyloxymethyltetraline (*110*) relative to exo-*111a* and endo-17-tosyloxymethyl-4,5-tetramethylene[2.2]paracyclophane (*111b*) is evidence both of participation of the paracyclophane system as neighboring group (*111a* and *111b* solvolyze faster than *110*), and of preferential exo carbonium ion formation (the exo-tosylate *111a* solvolyzes seven times faster than the endo-tosylate *111b*).

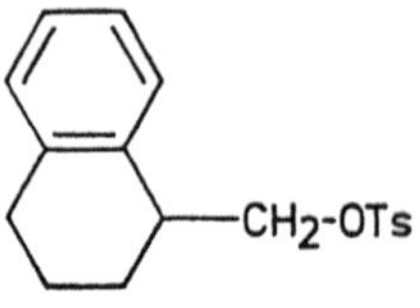

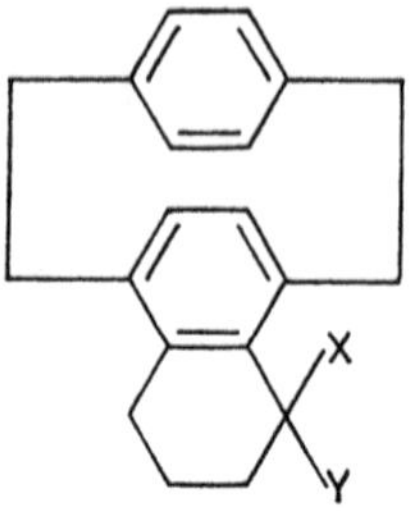

110

exo-*111a*: X = H; Y = CH_2OTs
endo-*111b*: X = CH_2OTs; Y = H

Product analysis shows that bridged ions of the type *112a* and *112b* may be involved and that their neutralization by the incoming nucleophile occurs exclusively at the exo carbon, a fact which is attributed to both steric and electronic effects.

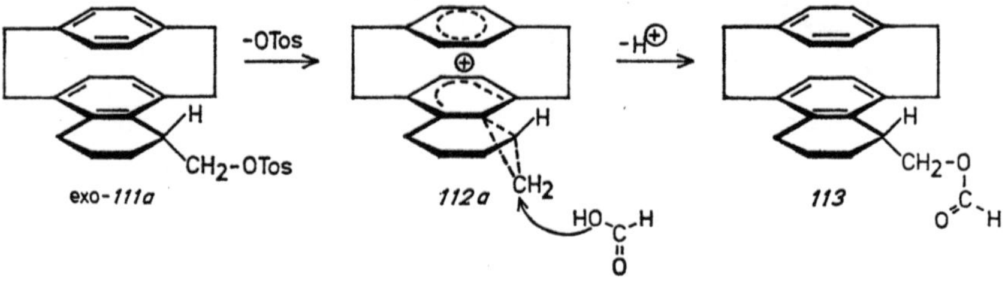

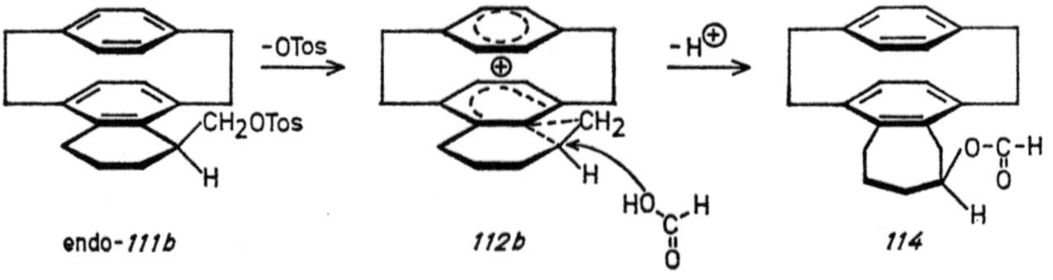

3.2. Photochemical Reactions of [2.2]Paracyclophane

Wasserman *et al.*[62,93] have shown that syn-[2.2](1,4)naphthalenophane (*34B*) is converted into the anti form in 70% yield when irradiated with light of wavelength 3500 Å in benzene for 10 days; this reaction also proceeds thermally at 250 °C. Photolysis of the pure anti isomer

34A in benzene affords small amounts of the syn product *34B* and relatively large amounts (25%) of dibenzoequinene (*115*) [94,95] with 70% unchanged starting material. *115* can be reconverted into anti-*34A* when heated in the solid phase at 200 °C.

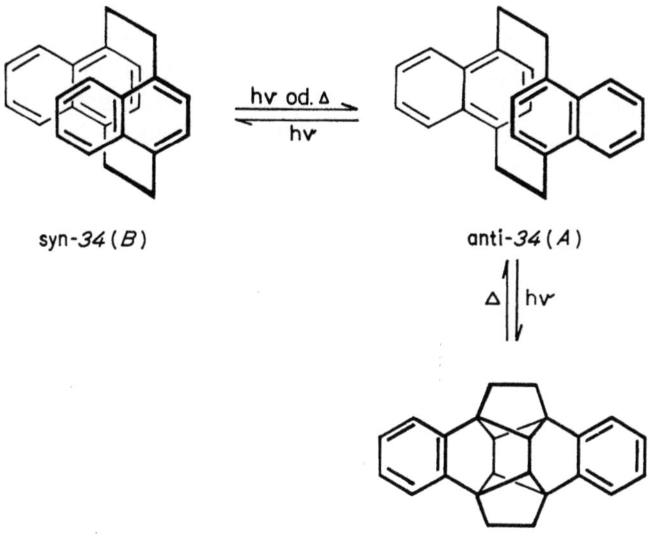

115

Irradiation of [2.2]paracyclophane, under different conditions (various solvents, light sources of different wavelength, addition of photosensitizers) always leads only to open-chain cleavage products of *2*. The counterpart of *115*, the polycyclic equinene (*116*), could not be detected [22]. Cram and Delton [96] even ruled out the intermediate occurrence of *116* analogs in the photo-racemization of a number of optically active nuclear- and bridge-substituted [2.2]paracyclophanes.

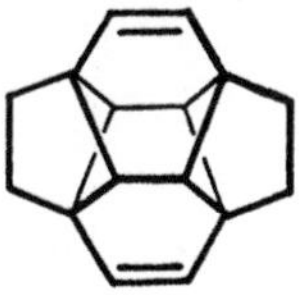

116

Golden [68] had already established that the orange-red, monoclinic [2.2](9,10)anthracenophane (*44*) is transformed into a colorless substance when exposed to sunlight, both in the solid phase and in chloroform solution. The colorless substance was assigned the structure *117*. On heating, *117* is converted into an apparently polymorphic form of *44*, which can be reconverted into the monoclinic form. According to recent investigations by Kaupp [69], *44* reacts uniformly without detectable fluorescence and with a quantum yield $\Phi = 0.30$ (CH_2Cl_2 or C_2H_5Br; 450, 365, 344, or 282 nm; N_2 or O_2) to give *117*, which is reconverted into *44* by 282 nm excitation with $\Phi = 0.6$. Since the sum of the quantum yields $\Phi_{forward} + \Phi_{reverse}$ is below the limiting value of 1, and neither oxygen nor ethyl bromide affect the quantum yields, both forward and reverse reactions must be assumed to have a common intermediate of type *118* having singlet character.

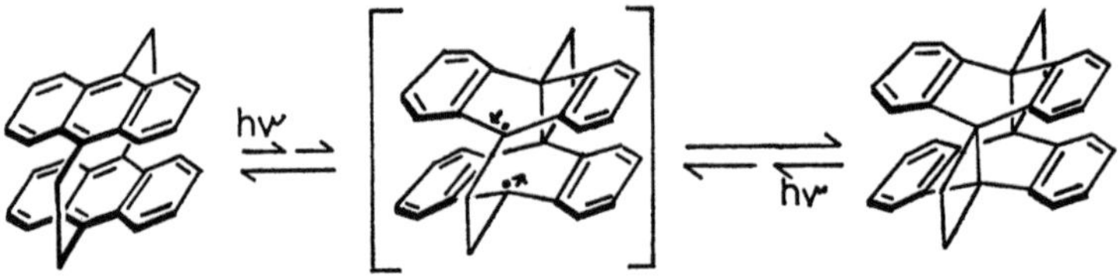

44 *118* *117*

When allowed to stand for several weeks in the sunlight in cyclohexane saturated with oxygen, [2]paracyclo[2](2,5)furanophane (*41 a*) generated [2.2.2]paracyclophane (*119*) together with some non-crystalline material. The fate of the furan component was not investigated [66b].

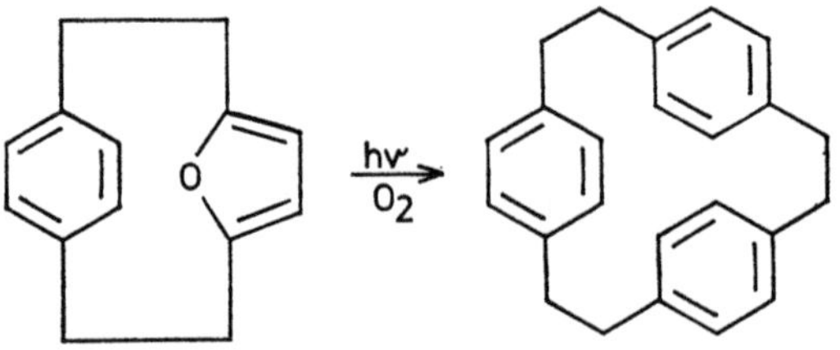

41a *119*

Wasserman *et al.* [97] irradiated a methanolic solution of *41 a* in the presence of oxygen and methylene blue for 18 h at 25 °C. After hydrogenation of the reaction mixture over palladium on charcoal for 24 h they were able to isolate three products: *120* (26%), *121* (11%), and 5% of a substance which was assigned structure *122* on the basis of spectroscopic and X-ray crystallographic data. The mechanism of formation of this polycycle probably involves an epoxide of type *123*

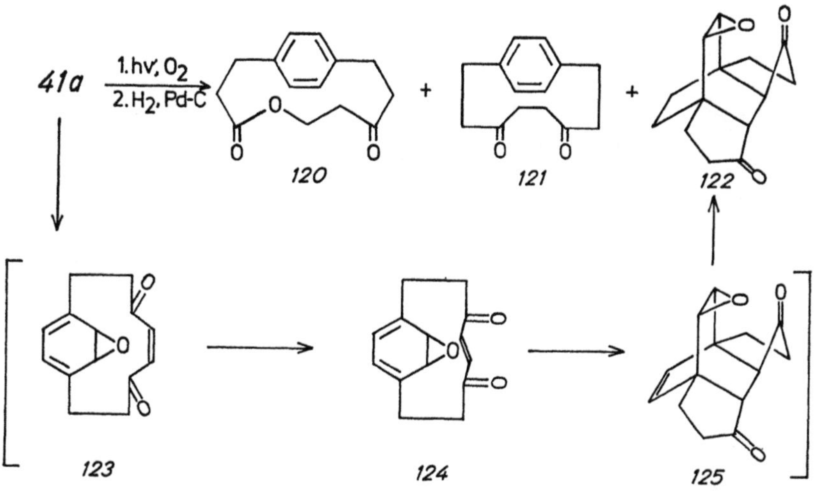

as intermediate; its isomerization to *124* and subsequent intramolecular Diels—Alder reaction then leads to *125*, which is hydrogenated to *122*. [2.2](2,5)Furanophane (*43*) reacts under similar conditions to give the diketone *126* [98].

126

Wasserman and Keehn [99] have also carried out the photosensitized auto-oxidation of anti-[2.2](1,4)naphthalenophane (*34A*). Irradiation of anti-*34* in methanol and simultaneous reaction with singlet oxygen affords the oxidation product *127* in 20% yield. The primary step in the reaction is assumed to be formation of a peroxide (*128*) whose geometry permits an intra-annular Diels—Alder reaction as second step; methanolysis then leads to *127* which was isolated.

The reaction of [2]paracyclo[2](1,4)naphthalenophane (*35*) with singlet oxygen in methanol [100] proceeds somewhat differently. In addition to unchanged starting material, the rearranged compound *129* and the cyclization product *130* were detected.

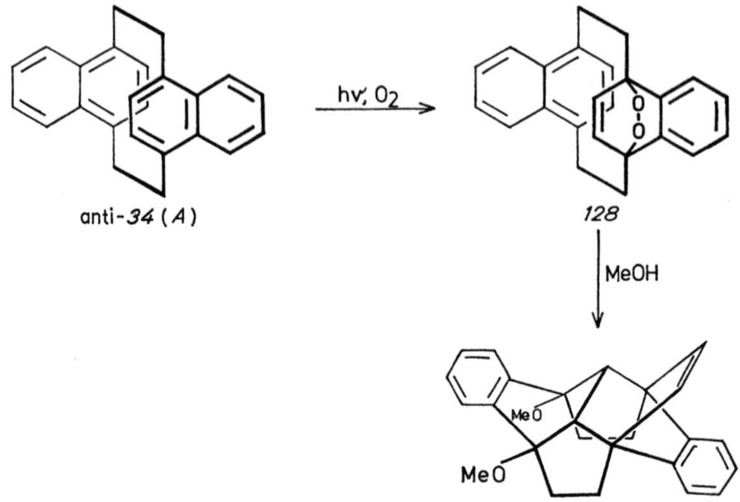

anti-*34* (*A*) *128* *127*

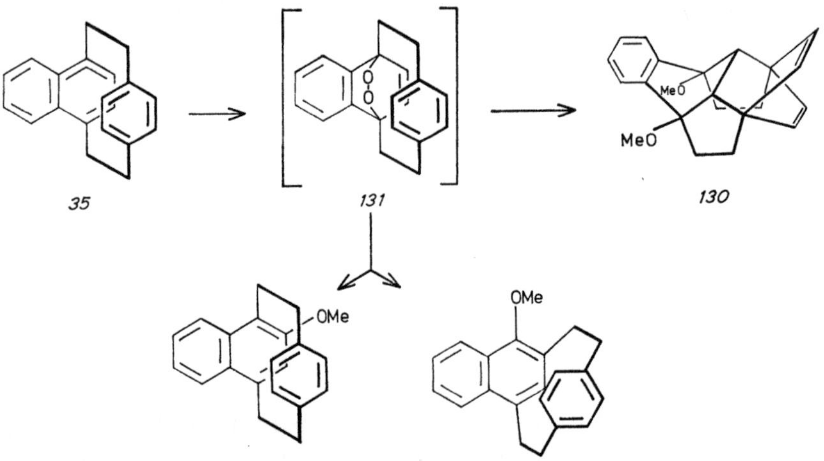

35 *131* *130* *132* *129*

The authors suggest a scheme for the reaction mechanism in which the initial step involves a 1,4-addition of oxygen to the substituted naphthalene nucleus (*35*→*131*). This is followed by methanolysis and intramolecular Diels—Alder addition giving the cyclization product *130*, which appears to be unstable compared to *127*.

The absorption pattern in the ^{1}H—NMR spectrum of the reaction product *129* corresponds in remarkable detail to the spectrum of 1-

methoxy-2,4-dimethylnaphthalene and is unlike that of 1,4-dimethyl-2-methoxynaphthalene, which corresponds to *132*.

Strong additional support for the assignment of the metapara-cyclophane structure *129* was obtained from dynamic NMR studies: The temperature dependence of the proton resonance of this compound is analogous to that of the parent [2.2]metaparacyclophane [3]. The multiplets observed for the protons of the p-phenyl nucleus gradually broaden with increasing temperature, disappear completely at 150 °C, and reappear at 180 °C as a "midway" peak. The isomer of *129*, the 13-methoxy[2]paracyclo[2](1,4)naphthalenophane (*132*), formation of which by the mechanism outlined above seems equally feasible, does not appear to occur.

3.3. Cycloadditions of [2.2]Paracyclophane and its Analogs

3.3.1. Diels—Alder Reactions

Intramolecular Diels—Alder reactions without prior 1,4-addition of oxygen (cf. previous section) have similarly been postulated for a number of [2.2]paracyclophane analogs. When [2](2,5)furano[2](1,4)naphthalen-ophane (*42*) is heated in excess dimethyl acetylenedicarboxylate at 100 °C, a polycyclic compound of structure *134* is formed. The mechanism of formation of *134* is most probably as follows [101]: the furan moiety reacts as active diene component in an intermolecular Diels—Alder reaction to give *135*. This is followed by further intramolecular 1,4-addition with the unsubstituted naphthalene ring as diene component to give the product *133*, which has been isolated.

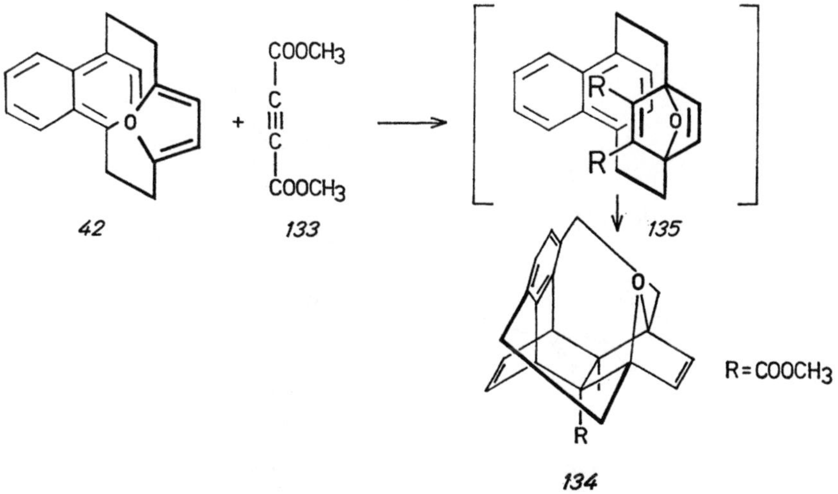

42 *133* *135* *134*

With [2](1,9)anthraceno[2](2,5)furanophane (*45*) the reaction takes a different course. Here, the intramolecular Diels—Alder reaction in the primary adduct *136* does not take place with the activated, substituted double bond as in the case of *42* and [2.2](2,5)furanophane (*43*), whose reactions with *133* have also been investigated [66b], but with the deactivated double bond functioning as dienophile. Spectroscopic findings indicate the structure *137* for the 1:2 adduct obtained when an excess of *133* is employed.

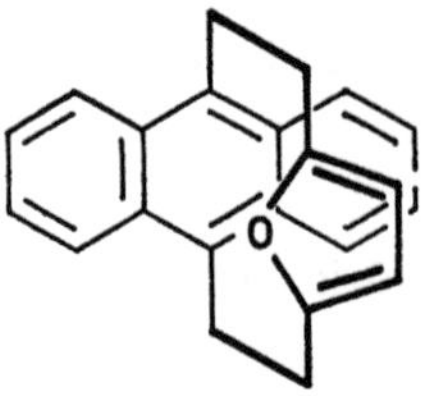

45

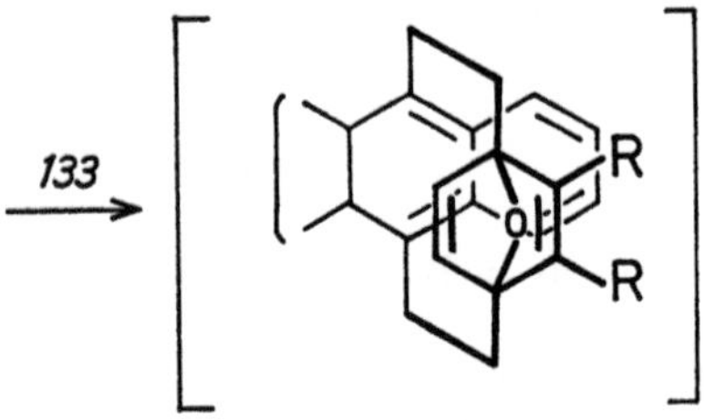

136

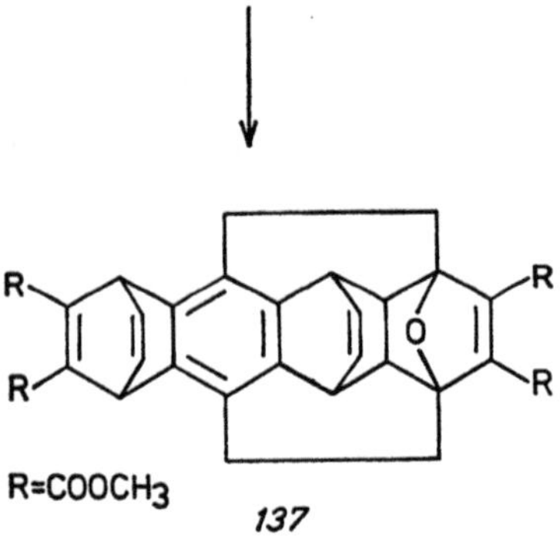

137

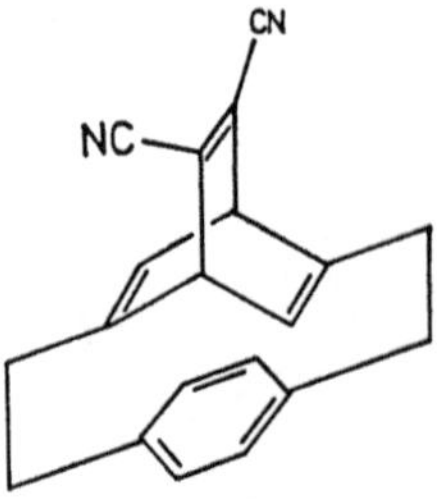

138

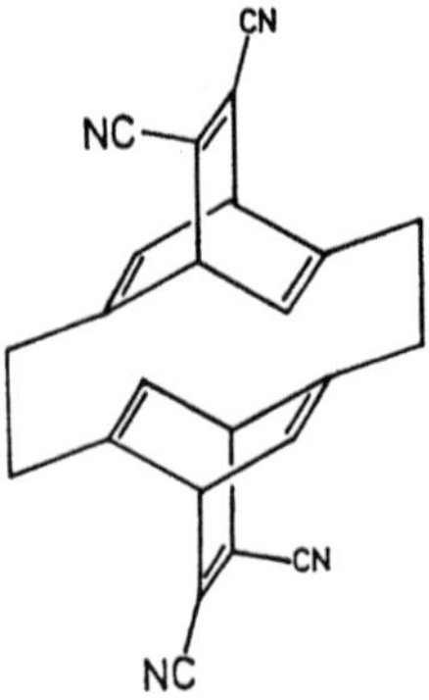

139

The ring strain in [2.2]paracyclophane lowers the activation barrier for uncatalyzed cycloaddition of dicyanoacetylene [102], so 1:1 and 2:1 adducts can be formed at relatively low temperatures. At 120 °C, a mixture of *138* and *139* is obtained; at 170 °C, only *139* is formed.

3.3.2. Cycloadditions with Dehydroaromatics

Brewer, Heaney, and Marples [103] were able to isolate the mono adduct *142* from the reaction of tetrafluorodehydrobenzene (*141*) with [2.2]paracyclophane. The 2:1 adducts *143* and *144* were obtained when an excess of pentafluorophenylmagnesium bromide was employed.

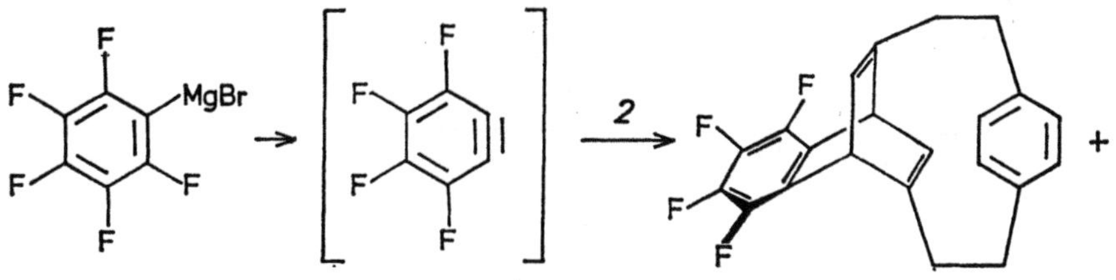

140 *141* *142*

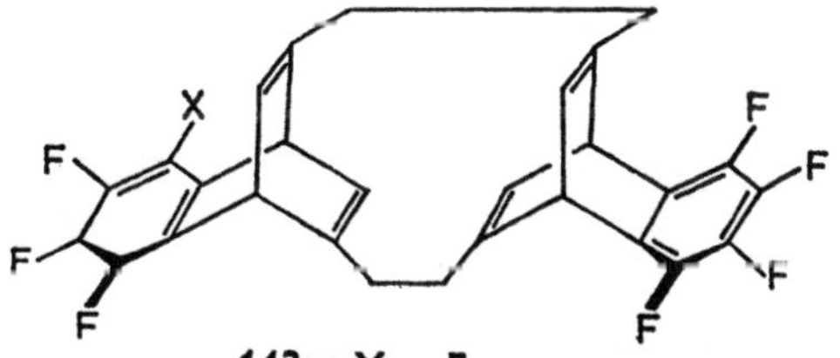

143a: X = F
144 : X = H

The intermediate occurrence of 4,5-dehydro[2.2]paracyclophane (*145*) was confirmed by Longone and Chipman [104] by trapping with anthracene. Excessive steric demands obviously cause deformation of the [2.2]paracyclophane moiety.

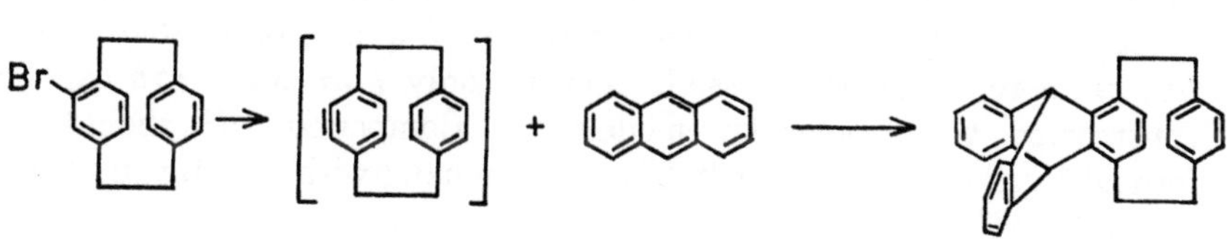

67 *145* *146* *147*

3.4. Oxidation and Reduction

3.4.1. Oxidation

Cram and Day [57] successfully synthesized a quinone of [2.2]paracyclophane by coupling the phenol *148* with diazotized sulphanilic acid to give *149*. Reduction of *149* gave the unstable aminophenol *150*; on oxidation with ferric sulphate *150* afforded the quinone *30* in 68% overall yield.

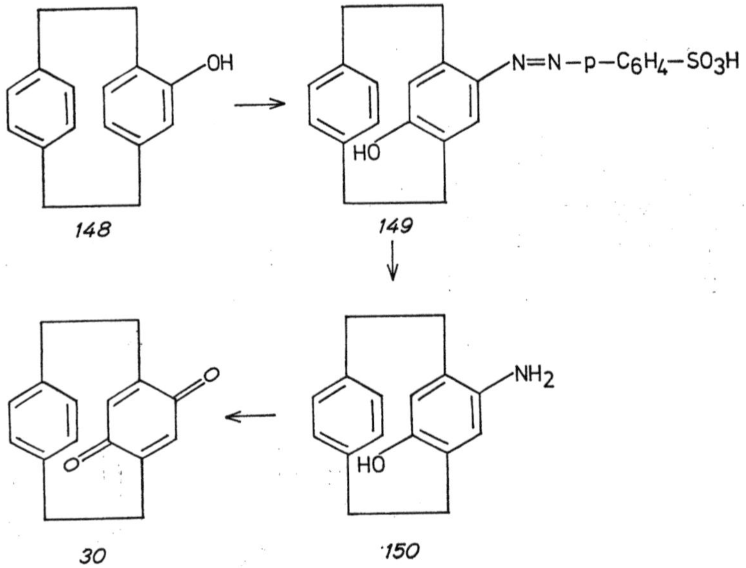

3.4.2. Reduction

The catalytic hydrogenation of [2.2]paracyclophane proceeds anomalously [37]. The first four mols of hydrogen are taken up much more rapidly than the second four. Reduction can therefore made to be terminated at the octahydro[2.2]paracyclophane (*151*). According to Cram, this rather unstable, waxlike substance, which can be isolated in up to 91% yield, probably has the structure *151a* or *151b*. The UV spectrum strongly supports the presence of a nonconjugated diene. Furthermore, of all the possible isomers, only *115a* and *115b* can be constructed with molecular models [105]. Compound *151* should be formed by the reduction of [2.2]paracyclophane with lithium in ethylamine [106,107].

Exhaustive hydrogenation (H_2/Pt in glacial acetic acid) provides the perhydro compound *152*, whose molecular architecture is indicated

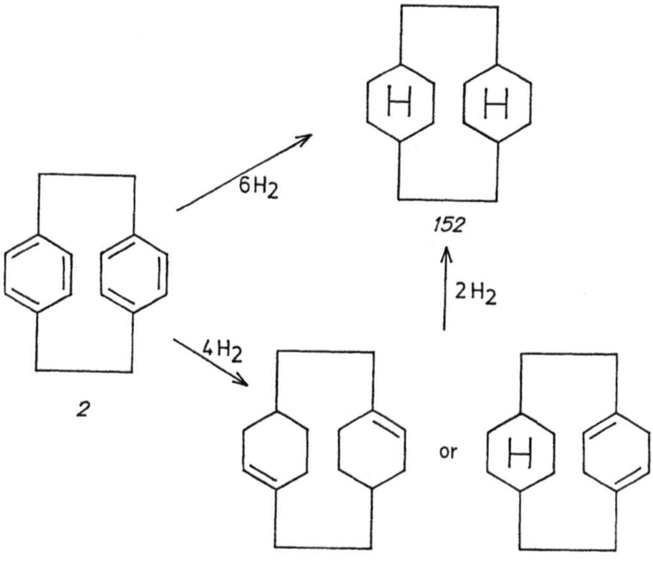

151a *151b*

by *152A*. It can readily be seen from molecular models that *152* is a compound whose molecular interior is sterically very overcrowed. The van der Waals radii overlap considerably and the bond angles are probably deformed [108]. Boyd *et al.*[9] have calculated the energetically most favorable conformations and strain energies of *152*.

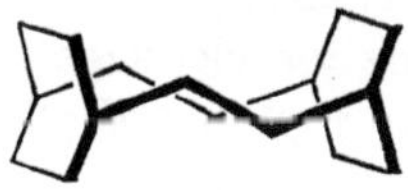

152 A

Jenny and Reiner [107] obtained the di- and tetrahydro compounds *153* and *154* by Birch reduction of [2.2]paracyclophane. On the basis of spectroscopic findings, the authors postulate the configuration *153aA*, for the dihydro compound, where the benzene rings are slightly distorted in a tublike fashion.

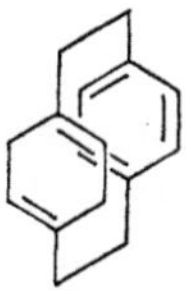

153a

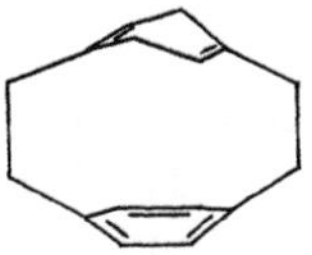

153a(A)

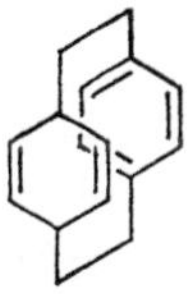

153b

In the case of the tetrahydro derivative *154* [108] it was demonstrated that there are two conceivable conformations that cannot be differentiated. Structures *154b* and *154c* can be ruled out on the basis of ^{1}H—NMR spectral data. Disproportionation of *154* to *153* and [2.2]paracyclophane occurs on heating to 150—160 °C in a closed system.

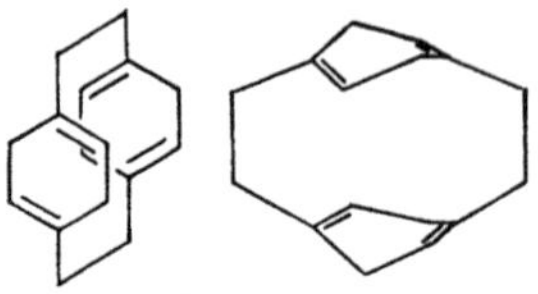

154a

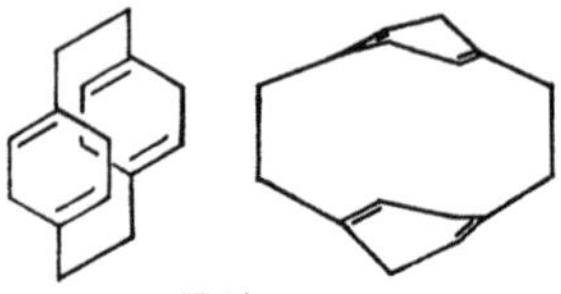

154b

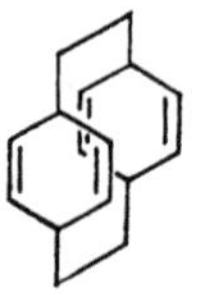

154c

3.5. Ring Fission Reactions

3.5.1. Thermal Isomerization and Racemization

The high strain energy of [2.2]paracyclophane (see Section 2.1.) facilitates ring-opening of the molecule via cleavage of the benzyl–benzyl bonds. Pyrolysis at 400 °C affords p,p'-dimethylbibenzyl (*155*) and p,p'-dimethylstilbene [109]. At 600 °C, p-xylylene (*156*) is formed; it polymerizes spontaneously to the linear poly-p-xylylene (*10*) on condensation [110].

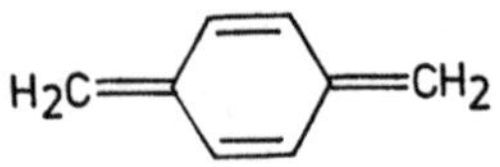

156

Reich and Cram [111,112] have reported the results of a series of experiments, which all indicate thermal cleavage of [2.2]paracyclophane to the p,p-dimethylenebibenzyl diradical (*157*). After heating [2.2]paracyclophane (*2*) at 250 °C in p-diisopropylbenzene, they isolated p,p-dimethylbibenzyl in 21% yield as the sole non-polymerizable product.

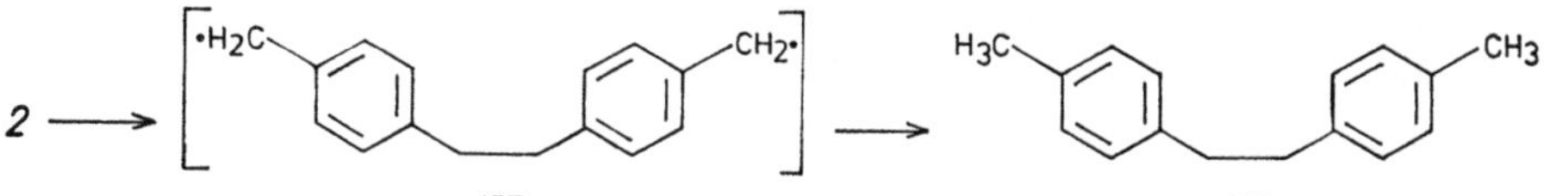

157 155

(−)-4-Carbomethoxy[2.2]paracyclophane (*70*) undergoes racemization without decomposition at 200 °C at a rate which shows very little dependence on solvent polarity; the values are nearly the same in dimethyl sulfone and tridecane. The rigidity of the ring system precludes racemization by a route which does not involve ring opening and subsequent recombination. The activation energy for the process was found to be $\Delta F^{\ddagger} \approx 38$ kcal/mol. Three alternative mechanisms were postulated: (i) homolytic cleavage of a benzyl-benzyl bond to the diradical *158*, whose recombination after rotation of benzene rings affords the racemic ester (Route *A*); (ii) cleavage of the methylene bridges to the p-xylylenes *156* and *159*, followed by recombination (p-xylylene mechanism) to racemic starting material (Route *B*); and (iii) intramolecular cycloaddition to the polycyclic equinenes *160* and *161*, followed by double cyclo-eliminations to racemic ester (Route *C*).

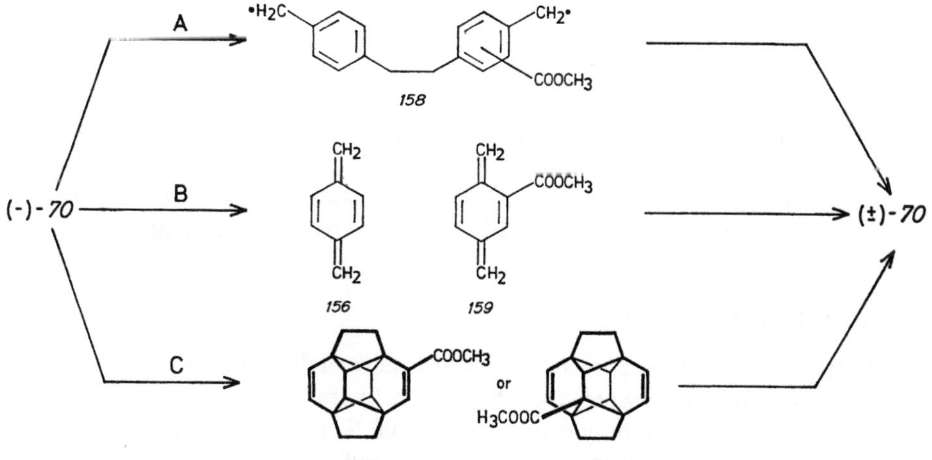

160 161

Investigations on the thermal isomerization of various disubstituted [2.2]paracyclophanes at 200 °C by Cram and Reich [111,112] showed that only the diradical mechanism (Route *A*) was consistent with experimental findings. Thermal isomerization starting from the pure pseudo-geminal and pseudo–meta isomers gave the same equilibrium mixture; the pseudo–ortho and pseudo–para isomers were similarly interconvertible:

A

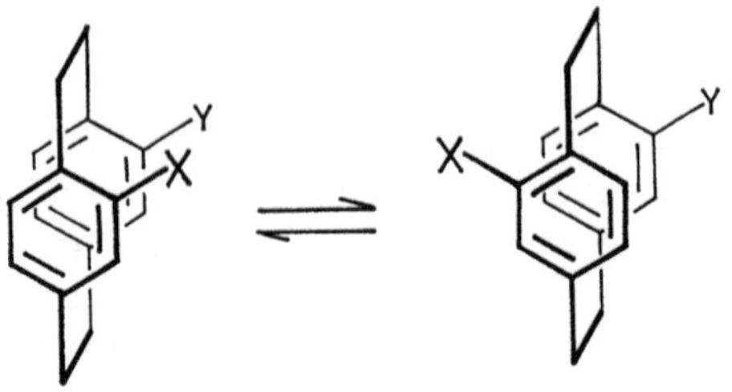

pseudo-gem.-*73* (X = Br, Y = $COCH_3$) pseudo-meta-*73*

B

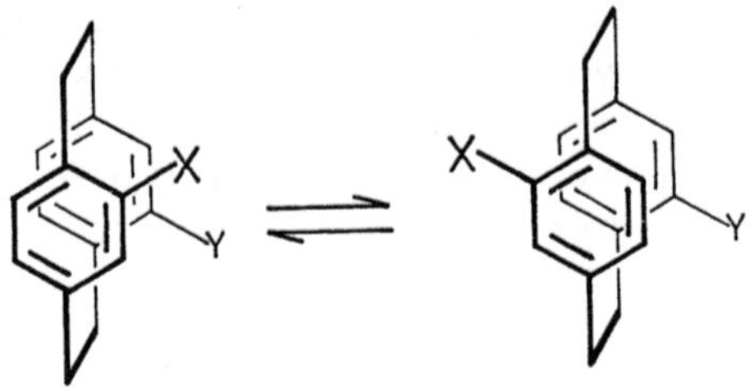

pseudo-ortho-*73* pseudo-para-*73*

Further isomerization from system *A* to system *B*, or of pseudo-geminal to meta- and pseudo–ortho to para–substituted products (leakage between sets of isomers), was not observed; this finding can be reconciled with a p-xylylene or polycyclic intermediate mechanism.

3.5.2. Ring Expansion

Further supporting evidence for the occurrence of diradicals was obtained by Reich and Cram when they heated [2.2]paracyclophane with either dimethyl maleate or fumarate esters at 200 °C for 40 h in the absence of air. The cis- and trans-2,3-dicarboxymethyl[4.2]paracyclophanes *162* and *163* were formed in about equal amounts, irrespective of the configuration of the olefin employed. Other similar reactions would also suggest a radical mechanism for this reaction; furthermore, a concerted addition of the olefinic double bond to *2*, or to the postulated intermediate diradical *157*, can be ruled out because of lack of stereospecificity of insertion.

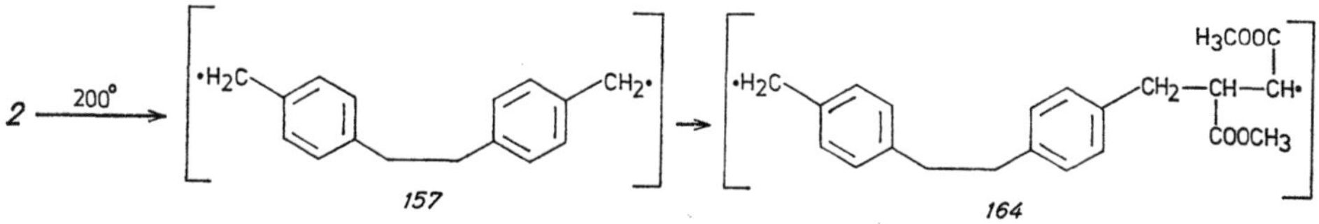

157 164

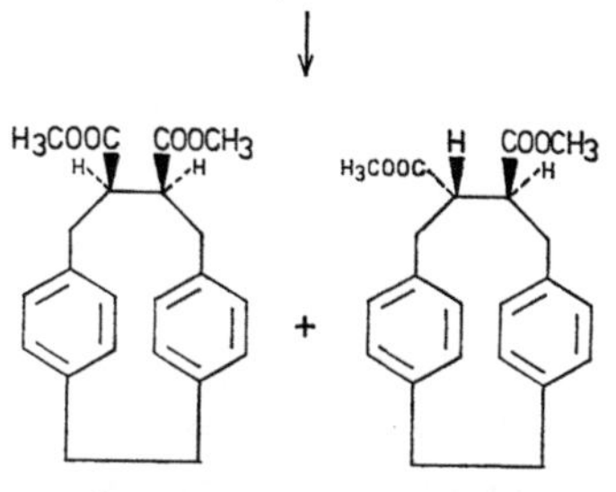

Yields on reaction with:	*cis-162*	*trans-163*
dimethyl maleate	30%	29%
dimethyl fumarate	34%	30%

cis-*162* trans-*163*

Radicals more stable than *157* should be formed on thermolysis of 1-vinyl[2.2]paracyclophane (*164*) and trans-β-carbomethoxy-1-vinyl-[2.2]paracyclophane (*165*) [113]. When heated at 165 °C, crystalline *164* gave the expanded ring cis-[4.2]paracyclophan-1-ene (*168*) in 90% yield; photolysis of *164* in methanol at room temperature also gave (13%) *168*. When heated at 100 °C in benzene, trans-*165* gave cis-3-carbomethoxy[2.2]paracyclophan-1-ene (*169*).

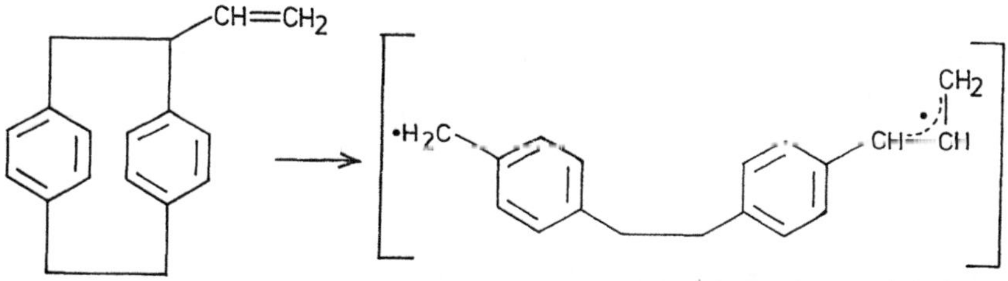

164 *166*

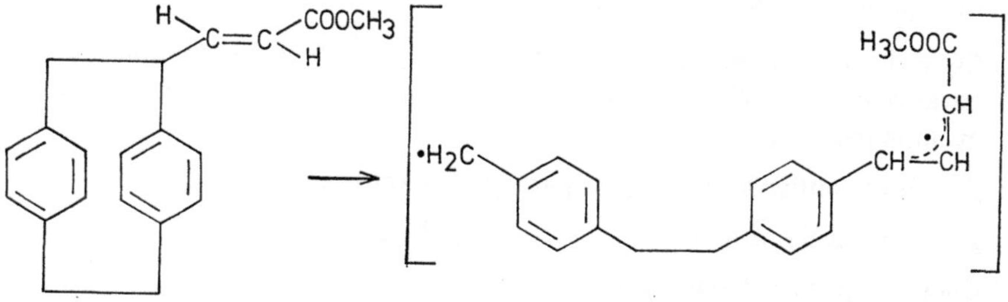

165 *167*

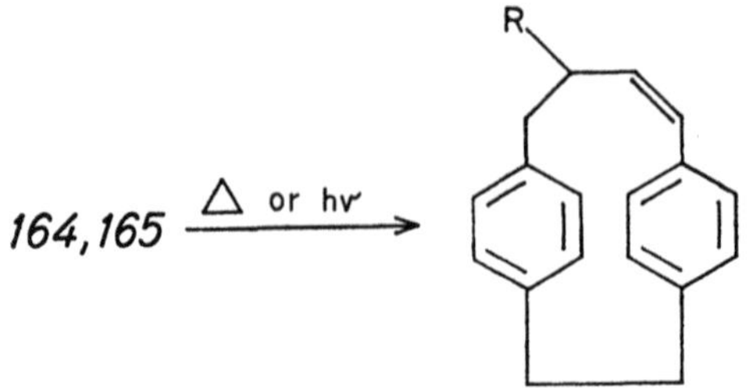

168: R = H
169: R = COOCH$_3$

Several products were obtained on heating *164* with dimethyl maleate or dimethyl fumarate: in addition to *168*, mixtures of the stereoisomers of 1-vinyl-2,3-biscarbomethoxy[2.2]paracyclophan-1-ene (*170*) and of the cis-4,5-biscarbomethoxy[2.2]paracyclophan-1-ene (*171*) were isolated. A similar product pattern was obtained on reaction of *165* with dimethyl maleate. The non-stereospecificity of these reactions as well as the kinetic data are in complete agreement with the previously postulated multi-step radical mechanism.

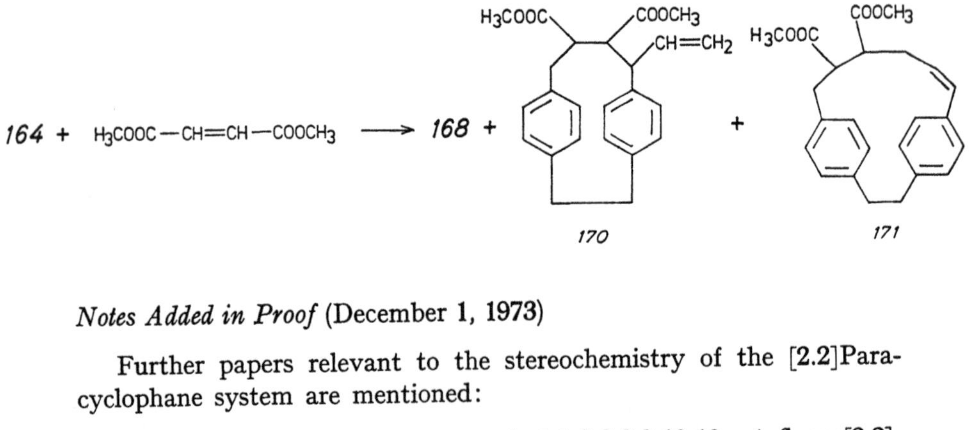

170 *171*

Notes Added in Proof (December 1, 1973)

Further papers relevant to the stereochemistry of the [2.2]Paracyclophane system are mentioned:

Crystal and molecular structure of 1,1,2,2,9,9,10,10-octofluoro[2.2]-paracyclophane (7) and a reinvestigation of the structure of [2.2]paracyclophane [114].

Absolute configuration of [2.2]paracyclophanes [115–117].

Anion radicals of a series of [2.2]paracyclophanes [118].

Chemiluminescent paracyclophanes [119].

Construction of additional bridges across [2.2]paracyclophanes [120].

[2.2]Paracyclophane-chinhydron [121].

[2.2]Anthracenophanes [122].

[2.2]Biphenylophanes and [2.2]naphthalinophane [123,124].

Pyridin analogs of [2.2]paracyclophane [125].

Layered [2.2]paracyclophanes [126–129].

Racemization and rearrangement mechanisms of chiral [2.2]para cyclophanes [130,131].

Stereochemical Elucidation of the Birch reduction product of [2.2]paracyclophanes [132].

Acknowledgments: We are greatly indebted to the Deutsche Forschungsgemeinschaft and the Fonds der Chemischen Industrie for support of our work mentioned in this paper.

4. References

1) Vögtle, F., Neumann, P.: Angew. Chem. *84*, 75 (1972); Angew. Chem. Intern. Ed. Engl. *11*, 73 (1972).
2) Vögtle, F., Neumann, P.: Chimia *26*, 64 (1972).
3) Brown, C. J., Farthing, A. C.: Nature *164*, 915 (1949).
4) Brown, C. J.: J. Chem. Soc. *1953*, 3265.
5) Lonsdale, K., Milledge, H. J., Krishna Rao, K. V.: Proc. Roy. Soc. (London) A *255*, 82 (1960).
6) Trueblood, K. N., Bernstein, J., Hope, H., cited in [6b] Cram, D. J., Cram, J. M.: Acc. Chem. Res. *4*, 204 (1971).
7) Boyd, R. H.: Tetrahedron *22*, 119 (1966).
8) Boyd, R. H.: J. Chem. Phys. *49*, 2574 (1968).
9) Shieh, C.-F., McNally, D., Boyd, R. H.: Tetrahedron *25*, 3653 (1969).
10) Cf. also Andrews, J. T. S., Westrum, Jr., E. F.: J. Phys. Chem. *74*, 2170 (1970).
11) Gantzel, P. K., Trueblood, K. N.: Acta Cryst. *18*, 958 (1965).
12) Anet, F. A. L., Brown, M. A.: J. Am. Chem. Soc. *91*, 2389 (1969).
13) Waugh, J. S., Fessenden, R W.: J. Am. Chem. Soc. *79*, 846 (1957).
14) Wilson, D. J., Boekelheide, V., Griffin, Jr., R. W.: J. Am. Chem. Soc. *82*, 6302 (1960).
15) Singer, L. A., Cram, D. J.: J. Am. Chem. Soc. *85*, 1080 (1963).
16) Cram, D. J., Dalton, C. K., Knox, G. R.: J. Am. Chem. Soc. *85*, 1088 (1963).
17) Cram, D. J., Helgeson, R. C.: J. Am. Chem. Soc. *88*, 3515 (1966).
18) Filler, R., Choe, E. W.: J. Am. Chem. Soc. *91*, 1862 (1969).
19) Cram, D. J., Steinberg, H.: J. Am. Chem. Soc. *73*, 5691 (1951).
20) Cram, D. J., Allinger, N. L., Steinberg, H.: J. Am. Chem. Soc. *76*, 6132 (1954).
21) Cram, D. J., Bauer, R. H., Allinger, N. L., Reeves, R. A., Wechter, W. J., Heilbronner, E.: J. Am. Chem. Soc. *81*, 5977 (1959).
22) Helgeson, R. C., Cram, D. J.: J. Am. Chem. Soc. *88*, 509 (1966).

23) a) Koutecky, J., Paldus, J.: Collection Czech. Chem. Commun. *27*, 599 (1962); Tetrahedron *19*, 201 (1963);
b) Paldus, J.: Collection Czech. Chem. Commun. *28*, 1110 (1963).
24) a) Ron, A., Schnepp, O.: J. Chem. Phys. *37*, 2540 (1962); *44*, 19 (1966);
b) Vala, Jr., M. T., Haebig, J., Rice, S. A.: J. Chem. Phys. *43*, 886 (1965);
c) Vala, Jr., M. T., Hillier, I. H., Rice, S. A., Jortner, J.: J. Chem. Phys. *44*, 23 (1966); Regarding the absorption spectrum of the [2.2]paracyclophane anion cf.
d) Hillier, I. H., Rice, S. A.: J. Chem. Phys. *45*, 4639 (1966).
25) Longworth, J. W., Bovey, F. A.: Bipolymers *4*, 1115 (1966).
26) For a detailed discussion of earlier studies see Smith, B. H.: Bridged aromatic compounds. New York—London: Academic Press 1964.
27) a) Gleiter, R.: Tetrahedron Letters *1969*, 4453; cf. also
b) Basu, S.: J. Chim. Phys. *62*, 827 (1965).
28) El-Sayed, M. A.: Nature *197*, 481 (1963).
29) Weissmann, S. I.: J. Am. Chem. Soc. *80*, 6462 (1958).
30) Ishitani, A., Nagakura, S.: Mol. Phys. *12*, 1 (1967).
31) Gerson, F., Martin, W. B.: J. Am. Chem. Soc. *91*, 1883 (1969).
32) Leo, A., Hansch, C., Elkins, D.: Chem. Rev. *71*, 525 (1971).
33) Dewhirst, K. C., Cram, D. J.: J. Am. Chem. Soc. *80*, 3115 (1958).
34) Chow, S. W., Pilato, L. A., Wheelwright, W. L.: J. Org. Chem. *35*, 20 (1970).
35) Hedaya, E., Kyle, L. M.: J. Org. Chem. *32*, 197 (1967).
36) Longone, D. T., Warren, C. L.: J. Am. Chem. Soc. *84*, 1507 (1962).
37) Cram, D. J., Allinger, N. L.: J. Am. Chem. Soc. *77*, 6289 (1955).
38) Cram, D. J., Wechter, R. J., Kierstead, R. W.: J. Am. Chem. Soc. *80*, 3126 (1958).
39) Cram, D. J., Reeves, R. H.: J. Am. Chem. Soc. *80*, 3094 (1958).
40) Lüttringhaus, A., Gralheer, H.: Liebigs Ann. Chem. *557*, 112 (1947).
41) It is expedient in the case of chiral derivatives of [2.2]paracyclophane to number the atoms independently of absolute configuration and in such a way that substituents have the lowest possible numbers. Absolute configuration is then indicated by adding the prefix R or S. Cf.
a) Falk, H., Reich—Rohrwig, P., Schlögl, K.: Tetrahedron *26*, 511 (1970).
42) Falk, H., Schlögl K.: Angew. Chem. *80*, 405 (1968); Angew. Chem. Intern. Ed. Engl. *7*, 383 (1968).
43) Falk, H., Schlögl, K.: Monatsh. Chem. *96*, 266 (1965).
44) Weigang, Jr., O. E., Nugent, M. J.: J. Am. Chem. Soc. *91*, 4555 (1969).
45) Nugent, M. J., Weigang, Jr., O. E.: J. Am. Chem. Soc *91*, 4556 (1969).
46) Reich, M. J., Cram, D. J.: J. Am. Chem. Soc. *91*, 3534 (1969).
47) a) Forrester, A. R., Ramasseul, R.: Chem. Commun. *1970*, 394;
b) J. Chem. Soc. [B] *1971*, 1638;
c) J. Chem. Soc. [B] *1971*, 1645.
48) a) Longone, D. T., Boettcher, F. P.: J. Am. Chem. Soc. *85*, 3436 (1963);
b) Longone, D. T., Chow, H. S.: J. Am. Chem. Soc. *86*, 3898 (1964);
c) Longone, D. T., Chow, H. S.: J. Am. Chem. Soc *92*, 994 (1970).
49) Otsubo, T., Mizogami, S., Sakata, Y., Misumi, S.: Tetrahedron Letters *1971*, 4803.
50) Longone, D. T., Reetz, M. T.: Chem. Commun. *1967*, 46.
51) Longone, D. T., Simanyi, L. H.: J. Org. Chem. *29*, 3245 (1964).
52) Hopf, H.: Angew. Chem. *84*, 471 (1972):
53) Nakazaki, M., Yamamoto, K., Tanaka, S.: Tetrahedron Letters *1971*, 341.
54) Filler, R., Miller, F. N.: Chem. Ind. (London) *1965*, 767.

55) Numerous references to through-space ^{1}H—^{19}F couplings have recently appeared in the literature;
a) Vögtle, F., Neumann, P.: Tetrahedron *26*, 5299 (1970);
b) Servis, K. L., Jerome, F. R.: J. Am. Chem. Soc. *93*, 1535 (1971);
c) Abushanab, E.: J. Am. Chem. Soc. *93*, 6532 (1971); and further literature cited therein.

56) For further information concerning the effective range of aromatically bound fluorine atoms cf.;
a) Nyburg, S. C., Szymansky, J. T.: Chem. Commun. *1968*, 669;
b) Vögtle, F.: Tetrahedron *25*, 3231 (1969);
c) Vögtle, F., Schunder, L.: Chem. Ber. *102*, 2677 (1969).

57) Cram, D. J., Day, A. C.: J. Org. Chem. *31*, 1227 (1966).

58) Coulter, C. L., Trueblood, K. N.: Acta Cryst. *16*, 667 (1963).

59) Keidel, F. A., Bauer, S. H.: J. Chem. Phys. *25*, 1218 (1956).

60) cis-Stilbene would certainly be a far more suitable compound for comparison; however, an X-ray structural analysis of this compound does not appear to have been carried out as yet. On the other hand, the C—aryl—C—alkenyl bond length in trans-stilbene is known:Robertson, J. M., Woodward, I.: Proc. Roy. Soc. (London) *A 162*, 568 (1937).

61) Brown, G. W., Sondheimer, F.: J. Am. Chem. Soc. *89*, 7116 (1967).

62) Wasserman, H. H., Keehn, P. M.: J. Am. Chem. Soc. *91*, 2374 (1969).

63) Toyoda, T., Otsubo, I., Otsubo, T., Sakata, Y., Misumi, S.: Tetrahedron Letters *1972*, 1731.

64) Vögtle, F., Neumann, P.: Tetrahedron *26*, 5847 (1970).

65) Wasserman, H. H., Keehn, P. M.: Tetrahedron Letters *1969*, 3227.

66) a) Cram, D. J., Knox, G. R.: J. Am. Chem. Soc. *83*, 2204 (1961);
b) Cram, D. J., Montgomery, C. S., Knox, G. R.: J. Am. Chem. Soc. *88*, 515 (1966).

67) Whitesides, G. M., Pawson, B. A., Cope, A. C.: J. Am. Chem. Soc *90*, 639 (1968).

68) Golden, J. H.: J. Chem. Soc. *1961*, 3741.

69) Kaupp, G.: Angew. Chem. *84*, 259 (1972).

70) Wynberg, H., Helder, R.: Tetrahedron Letters *1971*, 4317.

71) Haenel, M., Staab, H. A.: Tetrahedron Letters *1970*, 3585.

72) Bruhin, J., Jenny, W.: Chimia *25*, 238, 308 (1971).

73) For the method of preparation see Vögtle, F., Neumann, P.: Synthesis *1973*, 85.

74) For the nomenclature of "multilayered" phanes cf. 64).

75) Hillier, I. H., Glass, L., Rice, S. A.: J. Am. Chem. Soc. *88*, 5063 (1966).

76) Merrifield, R. E., Phillips, W. D.: J. Am. Chem. Soc. *80*, 2778 (1958).

77) Hubert, A. J.: J. Chem. Soc. [C] *1967*, 13.

78) Multi-bridged cyclophanes have been reviewed by Vögtle, F.: Chemiker-Ztg. *95*, 668 (1971).

79) Otsubo, T., Mizogami, S., Sakata, Y., Misumi, S.: Chem. Commun. *1971*, 678.

80) Mizogami, S., Otsubo, T., Sakata, Y., Misumi, S.: Tetrahedron Letters *1971*, 2791.

81) Fletcher, J. R., Sutherland, I. O.: Chem. Commun. *1969*, 1504.

82) Gault, I., Price, B. J., Sutherland, I. O.: Chem. Commun. *1967*, 540.

83) a) Reich, H. J., Cram, D. J.: J. Am. Chem. Soc. *90*, 1365 (1968);
b) J. Am. Chem. Soc. *91*, 3505 (1969).

84) For a detailed energetic study, the transition state must also be taken into consideration.

85) Reich, H. J., Cram, D. J.: J. Am. Chem. Soc. *91*, 3527 (1969).
86) It would be interesting to consider these mechanisms from the point of view of hard and soft acids and bases.
87) Forrester, A. R., Thomson, R. H.: J. Chem. Soc [C] *1966*, 1844.
88) a) Singler, R. E., Helgeson, R. C., Cram, D. J.: J. Am. Chem. Soc. *92*, 7625 (1970);
b) Singler, R. E., Cram, D. J.: ibid. *93*, 4443 (1971).
89) Cram, D. J., Singer, L. A.: J. Am. Chem. Soc. *85*, 1075 (1963).
90) Cram, D. J., Harris Jr., F. L.: J. Am. Chem. Soc. *89*, 4642 (1967).
91) a) Nugent, M. J., Vigo, T. L.: J. Am. Chem. Soc. *91*, 5483 (1969); cf. also b) Guest, A., Nugent, M. J.: Abstract of papers, 160th ACS National Meeting Chicago, Illinois, September 14—18, 1970.
92) For the preparation and properties of other compounds of type *111* see
a) Nugent, M. J.: Chem. Commun. *1967*, 1160;
b) Nugent, M. J., Vigo, T. L.: J. Org. Chem. *34*, 2203 (1969).
93) McBride, J. M., Keehn, P. M., Wasserman, H. H.: Tetrahedron Letters *1969*, 4147.
94) Wasserman, H. H., Keehn, P. M.: J. Am. Chem. Soc. *89*, 2770 (1967).
95) Fratini, A. V.: J. Am. Chem. Soc. *90*, 1688 (1968).
96) Delton, M. H., Cram, D. J.: J. Am. Chem. Soc. *92*, 7623 (1970).
97) Wasserman, H. H., Doumaux Jr., A. R., Davis, R. E.: J. Am. Chem. Soc. *88*, 4517 (1966).
98) Wasserman, H. H., Doumaux, Jr., A. R.: J. Am. Chem. Soc. *84*, 4611 (1962).
99) Wasserman, H. H., Keehn, P. M.: J. Am. Chem. Soc. *88*, 4522 (1966).
100) Wasserman, H. H., Keehn, P. M.: J. Am. Chem. Soc. *94*, 298 (1972).
101) Wasserman, H. H., Kitzing, R.: Tetrahedron Letters *1969*, 3343.
102) Ciganek, E.: Tetrahedron Letters *1967*, 3321.
103) Brewer, J. P. N., Heaney, H., Marples, B. A.: Tetrahedron *25*, 243 (1969).
104) Longone, D. T., Chipman, G. R.: Chem. Commun. *1969*, 1358.
105) Cram, D. J.: Rec. Chem. Progr. *20*, 71 (1959).
106) Cram, D. J., Bauer, R. H.: cited in [105].
107) Cf. Jenny, W., Reiner, J.: Chimia *24*, 69 (1970).
108) The preparation of the tetrahydro compound has also been reported by Marshall, J. L., Folsom, T. K.: Abstract of papers, 160th ACS Meeting, Chicago, Illinois, September 14—18, 1970.
109) Schaefgen, J. R.: J. Polymer. Sci. *15*, 203 (1955).
110) Gorham, W. F.: Am. Chem. Soc., Div. Polymer Chem., Reprint *6*, 73 (1965); C. A. 65: 20222 f.
111) Reich, H. J., Cram, D. J.: J. Am. Chem. Soc. *89*, 3078 (1967).
112) Reich, H. J., Cram, D. J.: J. Am. Chem. Soc. *91*, 3517 (1969).
113) Delton, M. H., Cram, D. J.: J. Am. Chem. Soc. *94*, 1669 (1972).
114) Hope, H., Bernstein, J., Trueblood, K. N.: Acta Cryst. B. *28*, 1733 (1972).
115) Eberhardt, H., Schlögl, K.: Liebigs Ann. Chem. *760*, 157 (1972).
116) Guest, A., Hoffman, P. H., Nugent, M. J.: J. Am. Chem. Soc. *94*, 4241 (1972).
117) Nugent, M. J., Guest, A.: J. Am. Chem. Soc. *94*, 4244 (1972).
118) Pearson, J. M., Williams, D. J., Levy, M.: J. Am. Chem. Soc. *93*, 5478 (1971).
119) Gundermann, K.-D., Röker, K.-D.: Angew. Chem. *85*, 451 (1973).
120) Truesdale, E. A., Cram, D. J.: J. Am. Chem. Soc. *95*, 5825 (1973).
121) Rebafka, W., Staab, H. A.: Angew. Chem. *85*, 831 (1973).
122) Iwama, A., Toyoda, T., Otsubo, T., Misumi, M.: Tetrahedron Letters *1973*, 1725.
123) Staab, H. A., Haenel, M.: Chem. Ber. *106*, 2190 (1973).

124) Haenel, M., Staab, H. A.: Chem. Ber. *106*, 2203 (1973).
125) Bruhin, J., Jenny, W.: Chimia *26*, 420 (1972).
126) Otsubo, T., Tozuka, Z., Mizogami, S., Sakata, Y., Misumi, S.: Tetrahedron Letters *1972*, 2927.
127) Nakazaki, M., Yamamoto, K., Tanaka, S.: Chem. Commun. *1972*, 433.
128) Nakazaki, M., Yamamoto, K., Ito, M.: Chem. Commun. *1972*, 433.
129) Kaneda, T., Ogawa, T., Misumi, S.: Tetrahedron Letters *1973*, 3373.
130) Delton, M. H., Cram, D. J.: J. Am. Chem. Soc. *94*, 2471 (1972).
131) Gilman, R. E., Delton, M. H., Cram, D. J.: J. Am. Chem. Soc. *94*, 2478 (1972).
132) Marshall, J. L., Song B.-H.: Abstract of papers, 166th ACS Meeting, August 26—31 (1973).

English version received May 7, 1973

www.ingramcontent.com/pod-product-compliance
Ingram Content Group UK Ltd.
Pitfield, Milton Keynes, MK11 3LW, UK
UKHW020239061126
16351UKWH00016B/53
* 9 7 8 3 6 6 2 1 5 5 5 9 2 *